It's another Quality Book from CGP

This book is for anyone doing Edexcel Modular GCSE Mathematics at Higher Level.

Whatever subject you're doing it's the same old story — there are lots of facts and you've just got to learn them. KS4 Maths is no different.

Happily this CGP book gives you all that important information as clearly and concisely as possible.

It's also got some daft bits in to try and make the whole experience at least vaguely entertaining for you.

What CGP is all about

Our sole aim here at CGP is to produce the highest quality books — carefully written, immaculately presented and dangerously close to being funny.

Then we work our socks off to get them out to you — at the cheapest possible prices.

Published by Coordination Group Publications Ltd.

Written by Richard Parsons.

Updated by Simon Little and Alison Palin.

With thanks to Sharon Keeley and Vicky Daniel for the proofreading.

ISBN: 978 1 84146 546 3

Groovy website: www.cgpbooks.co.uk
Printed by Elanders Hindson Ltd, Newcastle upon Tyne

Contents

Probability

This is nobody's favourite subject for sure, I've never really spoken to anyone who's said they do like it (not for long anyway). Although it does seem a bit of a "Black Art" to most people, it's not as bad as you might think, but YOU MUST LEARN THE BASIC FACTS.

All Probabilities are between 0 and 1

A probability of ZERO means it will NEVER HAPPEN,
A probability of ONE means it DEFINITELY WILL.

You can't have a probability bigger than 1.

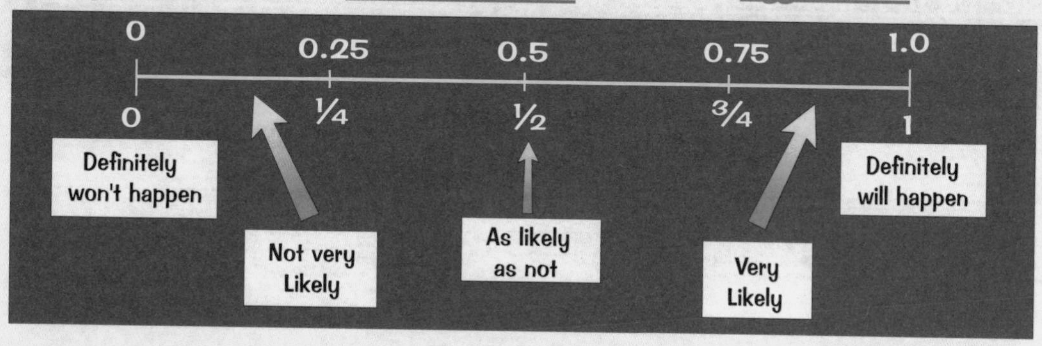

You should be able to put the probability of any event happening on this scale of 0 to 1.

Three Important Details

1) PROBABILITIES SHOULD BE GIVEN as either A FRACTION (¼), or A DECIMAL (0.25) or A PERCENTAGE (25%)
2) THE NOTATION : "P(x) = ½" SHOULD BE READ AS: "The probability of event X happening is ½"
3) PROBABILITIES ALWAYS ADD UP TO 1. This is essential for finding the probability of the other outcome. e.g. If P(pass) = ¼, then P(fail) = ¾

Listing All Outcomes: 2 Coins, Dice, Spinners

A simple question you might get is to list all the possible results from tossing two coins or two spinners or a dice and a spinner, etc. Whatever it is, it'll be very similar to these, so LEARN THEM:

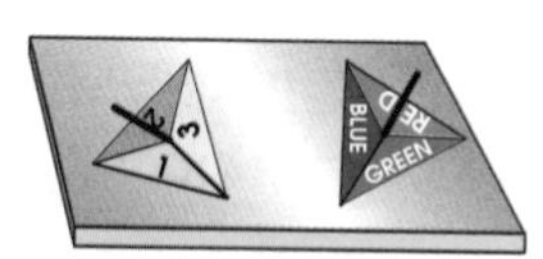

The possible outcomes from TOSSING TWO COINS are:

Head	Head	H H
Head	Tail	H T
Tail	Head	T H
Tail	Tail	T T

From TWO SPINNERS with 3 sides:

BLUE + 1	RED + 1	GREEN + 1
BLUE + 2	RED + 2	GREEN + 2
BLUE + 3	RED + 3	GREEN + 3

Try and list the possible outcomes METHODICALLY — to make sure you get them ALL.

Probability

This is where most people start getting into trouble, and d'you know why?
I'll tell you — it's because they don't know the three simple steps and the two rules to apply:

The Steps

1) Always break down a complicated-looking probability question into A SEQUENCE of SEPARATE SINGLE EVENTS.
2) Find the probability of EACH of these SEPARATE SINGLE EVENTS.
3) Apply the AND/OR rule:

The Rules:

1) The AND Rule:

$$P(A \text{ and } B) = P(A) \times P(B)$$

Which means:

The probability of Event A AND Event B BOTH happening is equal to the two separate probabilities MULTIPLIED together.

(strictly speaking, the two events have to be INDEPENDENT. All that means is that one event happening does not in any way affect the other one happening. Contrast this with mutually exclusive below.)

2) The OR Rule:

$$P(A \text{ or } B) = P(A) + P(B)$$

Which means:

The probability of Either Event A OR Event B happening is equal to the two separate probabilities ADDED together.

(Strictly speaking, the two events have to be MUTUALLY EXCLUSIVE which means that if one event happens, the other one can't happen.)

The way to remember this is that it's the wrong way round — i.e. you'd want the AND to go with the + but it doesn't: It's "AND with ×" and "OR with +".

Example

"Find the probability of picking two kings from a pack of cards (assuming you don't replace the first card picked)."

ANSWER:

1) SPLIT this into TWO SEPARATE EVENTS — i.e. picking the first king and then picking the second king.
2) Find the SEPARATE probabilities of these two separate events:

   P(1st king) = $^4/_{52}$ P(2nd king) = $^3/_{51}$ (— note the change from 52 to 51)
3) Apply the AND/OR Rule: BOTH events must happen, so it's the AND Rule:

   so multiply the two separate probabilities: $^4/_{52} \times ^3/_{51} = ^1/_{221}$

The Acid Test:

LEARN the Three Simple Steps for multiple events, and the AND/OR Rule.

1) Find the probability of picking from a pack of cards (without replacement):
   a) 2 queens plus the ace of spades. b) A pair of Jacks, Queens or Kings

Relative Frequency

This isn't the number of times your granny comes to visit.
It's a way of working out probabilities.

Fair or Biased?

The probability of rolling a three on a dice is $\frac{1}{6}$ — you know that each of the 6 numbers on a dice is equally likely to be rolled, and there's only 1 three.

BUT this only works if it's a fair dice. If the dice is a bit wonky (the technical term is "biased") then each number won't have an equal chance of being rolled. That's where Relative Frequency comes in — you can use it to work out probabilities when things might be wonky.

Do the Experiment Again and Again and Again and Again

You need to do an experiment over and over again and then do a quick calculation.
(Remember, an experiment could just mean rolling a dice.)
Usually the results of these experiments will be written in a table.

The Formula for Relative Frequency

$$\text{Probability of something happening} = \frac{\text{Number of times it has happened}}{\text{Number of times you tried}}$$

You can work out the relative frequency as a fraction but usually decimals are best for comparing relative frequencies.

The important thing to remember is:

The more times you do the experiment, the more accurate the probability will be.

Example:

So, back to the wonky dice. What is the probability of rolling a three?

Number of Times the dice was rolled	10	20	50	100
Number of threes rolled	3	9	19	36
Relative frequency	$\frac{3}{10}=0.3$	$\frac{9}{20}=0.45$	$\frac{19}{50}=0.38$	$\frac{36}{100}=0.36$

So, what's the probability? We've got 4 possible answers, but the best is the one worked out using the highest number of dice rolls.
This makes the probability of rolling a three on this dice 0.36.

And since for a fair, unbiased dice, the probability of rolling a three is $\frac{1}{6}$ (about 0.17), then our dice is probably biased.

The Acid Test:

LEARN the formula for calculating RELATIVE FREQUENCY

1) Bill picks a card at random out of a pack and then replaces the card. He does this 100 times, and picks a total of 13 aces. Do you think the pack is biased? Why?

Probability — Tree Diagrams

General Tree Diagram

Tree Diagrams are all pretty much the same, so it's a pretty darned good idea to learn these basic details (which apply to ALL tree diagrams) — ready for the one in the Exam.

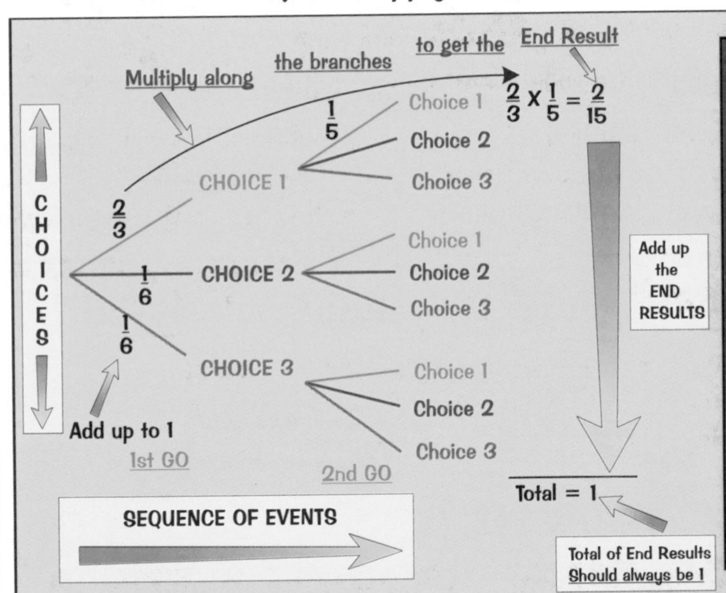

1) Always MULTIPLY ALONG THE BRANCHES (as shown) to get the END RESULTS.
2) On any set of branches which all meet at a point, the numbers must always ADD UP TO 1.
3) Check that your diagram is correct by making sure the End Results ADD UP TO ONE.
4) To answer any question, simply ADD UP THE RELEVANT END RESULTS (see below).

A likely Tree Diagram Question

EXAMPLE: "A box contains 5 red disks and 3 green disks. Two disks are taken without replacement. Draw a tree diagram and hence find the probability that both disks are the same colour."

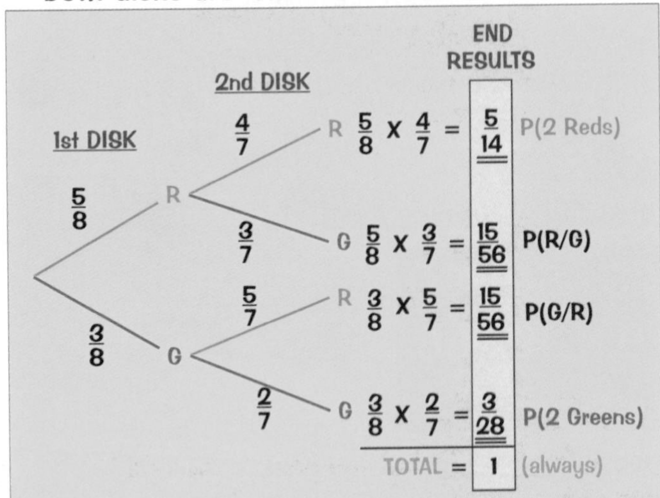

Once the tree diagram is drawn all you then need to do to answer the question is simply select the RELEVANT END RESULTS and then ADD THEM TOGETHER:

2 REDS (5/14)
2 GREENS (3/28)

$$\frac{5}{14} + \frac{3}{28} = \frac{13}{28}$$

If you can, use a calculator for this.

The Acid Test:

LEARN the GENERAL DIAGRAM for Tree Diagrams and the 4 points that go with them.

1) O.K. let's see what you've learnt shall we: TURN OVER AND WRITE DOWN EVERYTHING YOU KNOW ABOUT TREE DIAGRAMS.
2) A bag contains 6 red tarantulas and 4 black tarantulas. If two girls each pluck out a tarantula at random, draw a tree diagram to find the probability that they get different coloured ones.

Probability — Tree Diagrams

Four Extra Details for the Tree Diagram method:

1) Always break up the question INTO A SEQUENCE OF SEPARATE EVENTS.

E.g. "3 coins are tossed together" – just split it into 3 separate events.
You need this sequence of events to be able to draw any sort of tree diagram.

2) DON'T FEEL you have to draw COMPLETE tree diagrams.

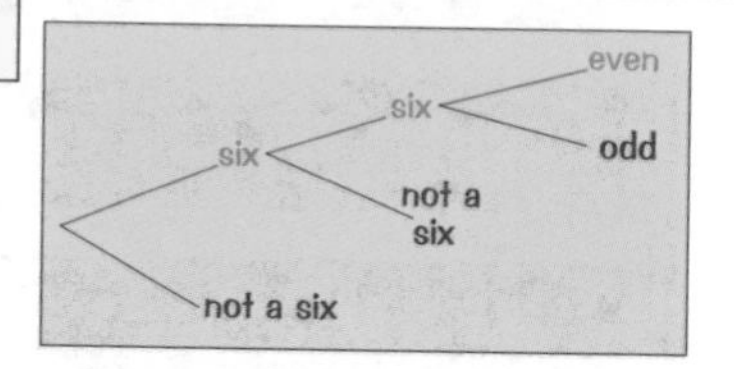

Learn to adapt them to what is required. E.g. "What is the chance of throwing a dice 3 times and getting 2 sixes followed by an even number?"
This diagram is all you need to get the answer: $\frac{1}{6}\times\frac{1}{6}\times\frac{1}{2}=\frac{1}{72}$

3) WATCH OUT for CONDITIONAL PROBABILITIES...

...where the fraction on each branch depends on what happened on the previous branch, e.g. bags of sweets, packs of cards etc, where the bottom number of the fractions also changes as items are removed. E.g. $^{11}/_{25}$ then $^{10}/_{24}$ etc.

4) With "AT LEAST" questions, it's always (1 – Prob of "the other outcome"):

For Example, "Find the probability of having AT LEAST one girl in 4 children"
There are in fact 15 different ways of having "AT LEAST one girl in 4 children" which would take a long time to work out, even with a tree diagram.
The clever trick you should know is this:
The prob of "AT LEAST something or other" is just (1 – prob of "the other outcome") which in this case is (1 – prob of "all 4 boys") = (1 – 1/16) = 15/16.

Example

"Herbert and his two chums, along with five of Herbert's doting aunties, have to squeeze onto the back seat of his father's Bentley, en route to Royal Ascot. Given that Herbert does not sit at either end, and that the seating order is otherwise random, find the probability of Herbert having his best chums either side of him."

The untrained probabilist wouldn't think of using a tree diagram here, but see how easy it is when you do. This is the tree diagram you'd draw:

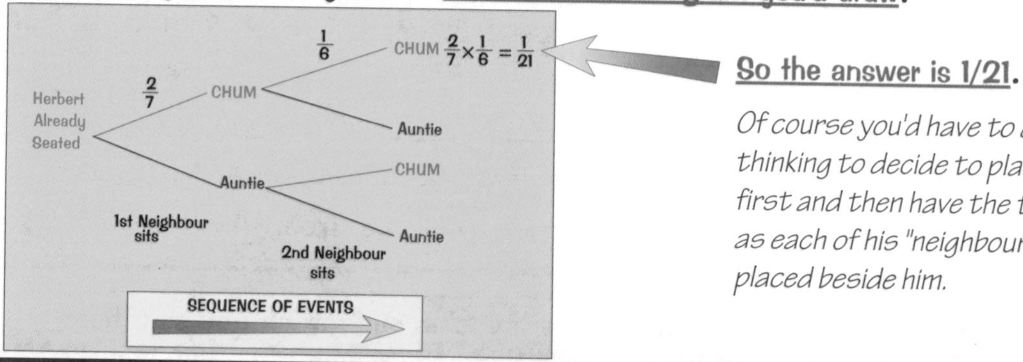

So the answer is 1/21.

Of course you'd have to do a bit of thinking to decide to place Herbert first and then have the two events as each of his "neighbours" are placed beside him.

The Acid Test:

LEARN THE WHOLE OF THIS PAGE. Then turn over and write down all the key points and the example too.

1) As it turned out, the Bentley could only seat 6 people across so the last two in had to sit on other people's laps. Find the probability that Herbert had his best chums either side and no doting Auntie on his lap (assuming Herbert wasn't one of the last 2).

Mean, Median, Mode and Range

If you don't manage to learn these 4 basic definitions then you'll be passing up on some of the easiest marks in the whole Exam. It can't be that difficult can it?

1) **MODE** = **MOST** common

2) **MEDIAN** = **MIDDLE** value

3) **MEAN** = **TOTAL** of items ÷ **NUMBER** of items

4) **RANGE** = How far from the smallest to the biggest

THE GOLDEN RULE

Mean, median and mode should be easy marks but even people who've gone to the incredible extent of learning them still manage to lose marks in the Exam because they don't do this one vital step:

Always REARRANGE the data in ASCENDING ORDER

(and check you have the same number of entries!)

Example

"Find the mean, median, mode and range of these numbers:"

2, 5, 3, 2, 6, -4, 0, 9, -3, 1, 6, 3, -2, 3 (14)

1) FIRST... rearrange them: -4, -3, -2, 0, 1, 2, 2, 3, 3, 3, 5, 6, 6, 9 (14)✓

2) MEAN = $\frac{\text{total}}{\text{number}} = \frac{-4-3-2+0+1+2+2+3+3+3+5+6+6+9}{14}$

= 31÷14=2.21

3) MEDIAN = the middle value (only when they are arranged in order of size, that is!).

When there are two middle numbers as in this case, then the median is HALFWAY BETWEEN THE TWO MIDDLE NUMBERS

-4, -3, -2, 0, 1, 2, 2, 3, 3, 3, 5, 6, 6, 9
← seven numbers this side ↑ seven numbers this side →
Median = 2.5

4) MODE = most common value, which is simply 3. (Or you can say "The modal value is 3")

5) RANGE = distance from lowest to highest value, i.e. from -4 up to 9, = 13

REMEMBER: Mode = most (emphasise the 'o' in each when you say them)
Median = mid (emphasise the m*d in each when you say them)
Mean is just the average, but it's mean 'cos you have to work it out.

The Acid Test: LEARN The Four Definitions and THE GOLDEN RULE...

..then turn this page over and write them down from memory. Then apply all that you have learnt to this set of data: 1, 3, 14, -5, 6, -12, 18, 7, 23, 10, -5, -14, 0, 25, 8.

Frequency Tables

Frequency Tables can either be done in rows or in columns of numbers and they can be quite confusing, but not if you learn these Eight key points:

Eight Key Points

1) ALL FREQUENCY TABLES ARE THE SAME.
2) The word FREQUENCY just means HOW MANY, so a frequency table is nothing more than a "How many in each group" table.
3) The FIRST ROW (or column) just gives the GROUP LABELS.
4) The SECOND ROW (or column) gives the ACTUAL DATA.
5) You have to WORK OUT A THIRD ROW (or column) yourself.
6) The MEAN is always found using: **3rd Row total ÷ 2nd Row Total**
7) The MEDIAN is found from the MIDDLE VALUE in the 2nd row.
8) The RANGE is found from the extremes of the first row.

Example

Here is a typical frequency table shown in both ROW FORM and COLUMN FORM:

No. of Sisters	Frequency
0	7
1	15
2	12
3	8
4	3
5	1
6	0

No. of Sisters	0	1	2	3	4	5	6
Frequency	7	15	12	8	3	1	0

Column Form

Row form

There's no real difference between these two forms and you could get either one in your Exam. Whichever you get, make sure you remember these THREE IMPORTANT FACTS:

1) THE 1ST ROW (or column) gives us the GROUP LABELS for the different categories: i.e. "no sisters", "one sister", "two sisters", etc.
2) THE 2ND ROW (or column) is the ACTUAL DATA and tells us HOW MANY (people) THERE ARE in each category i.e. 7 people had "no sisters", 15 people had "one sister", etc.
3) BUT YOU SHOULD SEE THE TABLE AS UNFINISHED, because it still needs A THIRD ROW (or column) and TWO TOTALS for the 2nd and 3rd rows, as shown on the next page:

Frequency Tables

This is what the two types of table look like when they're completed:

No. of sisters	0	1	2	3	4	5	6	totals	
Frequency	7	15	12	8	3	1	0	46	(People asked)
No. x Frequency	0	15	24	24	12	5	0	80	(Sisters)

No. of Sisters	Frequency	No. x Frequency
0	7	0
1	15	15
2	12	24
3	8	24
4	3	12
5	1	5
6	0	0
TOTALS	46	80
	(People asked)	(Sisters)

"Where does the third row come from?" ....I hear you cry!

THE THIRD ROW (or column) is ALWAYS obtained by MULTIPLYING the numbers FROM THE FIRST 2 ROWS (or columns).

THIRD ROW = 1ST ROW × 2ND ROW

Once the table is complete, you can easily find the MEAN, MEDIAN, MODE AND RANGE (see P.6) which is what they usually demand in the Exam:

Mean, Median, Mode and Range:

This is easy enough if you learn it. If you don't, you'll drown in a sea of numbers.

1) MEAN = $\frac{\text{3rd Row Total}}{\text{2nd Row Total}}$ = $\frac{80}{46}$ = 1.74 (Sisters per person)

2) MEDIAN: — imagine the original data SET OUT IN ASCENDING ORDER:

0000000 111111111111111 222222222222 33333333 444 5
↑

and the median is just the middle which is here between the 23rd and 24th digits. So for this data THE MEDIAN IS 2.

(Of course, when you get slick at this you can easily find the position of the middle value straight from the table)

3) The MODE is very easy – it's just THE GROUP WITH THE MOST ENTRIES: i.e 1

4) The RANGE is 5 – 0 = 5 The 2nd row tells us there are people with anything from "no sisters" right up to "five sisters" (but not 6 sisters). (Always give it as a single number)

The Acid Test:

LEARN the 8 RULES for Frequency Tables, then turn over and WRITE THEM DOWN to see what you know.

Using the methods you have just learned and this frequency table, find the MEAN, MEDIAN, MODE and RANGE of the no. of phones that people have.

No. of Phones	0	1	2	3	4	5	6
Frequency	1	25	53	34	22	5	1

Grouped Frequency Tables

These are a bit trickier. The table below shows the distribution of weights of 60 school kids:

Weight (kg)	$30 \le w < 40$	$40 \le w < 50$	$50 \le w < 60$	$60 \le w < 70$	$70 \le w < 80$
Frequency	8	16	18	12	6

What does $30 \le w < 40$ mean?

Don't get confused by the notation used for the intervals.

1) the $\le$ symbol means w can be greater than or equal to 30
2) the < symbol means w must be less than 40 (but not equal to it)

So a value of 30 will go in this group, but a value of 40 will have to go in the next group up: $40 \le w < 50$.

"Estimating" The Mean using Mid-Interval Values

Just like with ordinary frequency tables you have to add extra rows and find totals to be able to work anything out. Also notice you can only "estimate" the mean from grouped data tables — you can't find it exactly unless you know all the original values.

1) Add a 3rd row and enter MID-INTERVAL VALUES for each group.
2) Add a 4th row and multiply FREQUENCY × MID-INTERVAL VALUE for each group.

Weight (kg)	$30 \le w < 40$	$40 \le w < 50$	$50 \le w < 60$	$60 \le w < 70$	$70 \le w < 80$	TOTALS
Frequency	8	16	18	12	6	60
Mid-Interval Value	35	45	55	65	75	—
Frequency × Mid-Interval Value	280	720	990	780	450	3220

1) ESTIMATING THE MEAN is then the usual thing of DIVIDING THE TOTALS:

$$\text{Mean} = \frac{\text{Overall Total (Final Row)}}{\text{Frequency Total (2nd Row)}} = \frac{3220}{60} = \underline{53.7}$$

2) THE MODE is still nice'n'easy: the modal group is $50 \le w < 60$ kg
3) THE MEDIAN can't be found exactly but you can say which group it's in. If all the data were put in order, the 30th/31st entries would be in the $50 \le w < 60$ kg group.

The Acid Test:

LEARN all the details on this page, then turn over and write down everything you've learned. Good clean fun.

1) Estimate the mean for this table:
2) Also state the modal group and the approximate value of the median.

Length L (cm)	$15.5 \le L < 16.5$	$16.5 \le L < 17.5$	$17.5 \le L < 18.5$	$18.5 \le L < 19.5$
Frequency	12	18	23	8

Cumulative Frequency

FOUR KEY POINTS

1) CUMULATIVE FREQUENCY just means ADDING IT UP AS YOU GO ALONG.
2) You have to ADD A THIRD ROW to the table — the RUNNING TOTAL of the 2nd row.
3) When plotting the graph, always plot points using the HIGHEST VALUE in each group (of row 1) with the value from row 3. i.e. plot 13 at 160, etc. (see below).
4) CUMULATIVE FREQUENCY is always plotted up the side of a graph, not across.

Example

Height (cm)	$140 \le x < 150$	$150 \le x < 160$	$160 \le x < 170$	$170 \le x < 180$	$180 \le x < 190$	$190 \le x < 200$	$200 \le x < 210$
Frequency	4	9	20	33	36	15	3
Cumulative Frequency	4 (AT 150)	13 (AT 160)	33 (AT 170)	66 (AT 180)	102 (AT 190)	117 (AT 200)	120 (AT 210)

The graph is plotted from these pairs: (150, 4) (160, 13) (170, 33) (180, 66) etc.

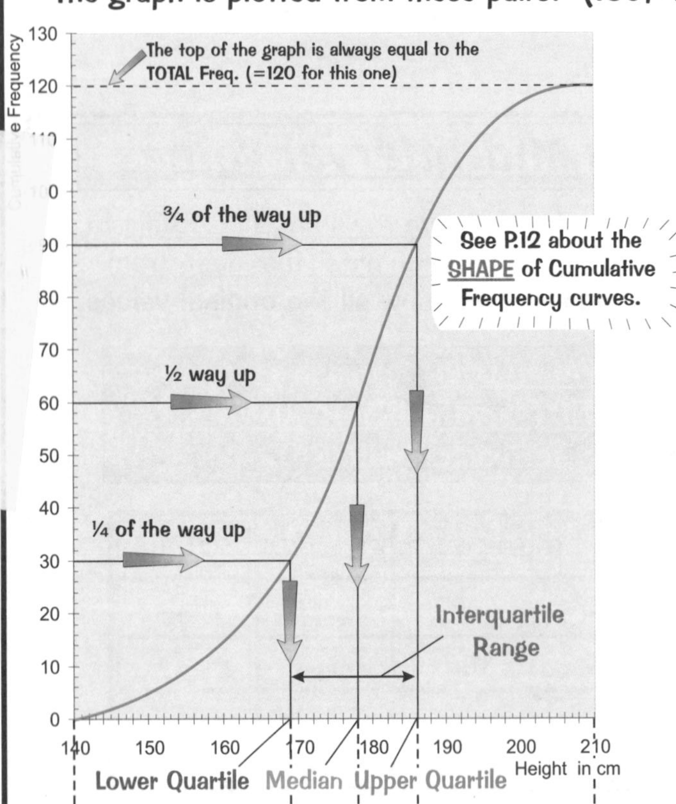

For a cumulative frequency curve there are THREE VITAL STATISTICS which you need to know how to find:

1) **MEDIAN**
Exactly halfway UP, then across, then down and read off the bottom scale.

2) **LOWER AND UPPER QUARTILES**
Exactly ¼ and ¾ UP the side, then across, then down and read off the bottom scale.

3) **THE INTERQUARTILE RANGE**
The distance on the bottom scale between the lower and upper quartiles.

So from the cumulative frequency curve for this data, we get these results:

MEDIAN = 178cm
LOWER QUARTILE = 169cm
UPPER QUARTILE = 186cm
INTERQUARTILE RANGE = 17cm (186-169)

A Box Plot shows the Interquartile Range as a Box

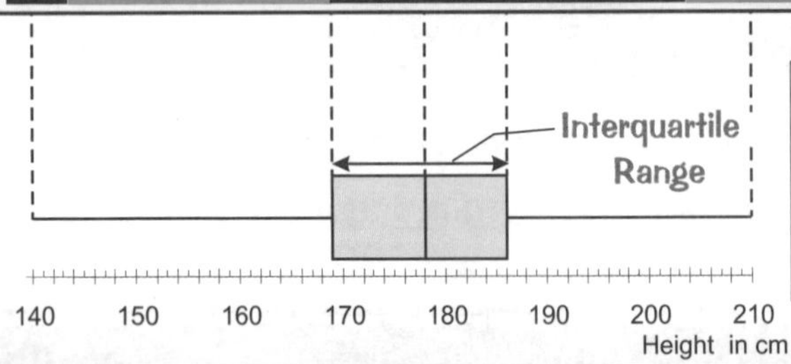

TO CREATE YOUR VERY OWN BOX PLOT:

1) Draw the scale along the bottom.
2) Draw a box the length of the interquartile range.
3) Draw a line down the box to show the median.
4) Draw "whiskers" up to the maximum and minimum.

(They're sometimes called "Box and Whisker diagrams".)

The Acid Test:

LEARN THIS PAGE, then cover it up and do these:

1) Complete this cumulative frequency table:
2) Draw the graph. Find the 3 Vital Statistics.
3) Draw the box plot under the graph.

No of fish	41 – 45	46 – 50	51 – 55	56 – 60	61 – 65	66 – 70	71 – 75
Frequency	2	7	17	25	19	8	2

Histograms and Frequency Density

Histograms

A histogram is just a bar chart where the bars can be of DIFFERENT widths.

This changes them from nice easy-to-understand diagrams into seemingly incomprehensible monsters, and yes, you've guessed it, that makes them a firm favourite with the Examiners.

In fact things aren't half as bad as that — but only if you LEARN THE THREE RULES:

1) It's not the height, but the AREA of each bar that matters.
2) Use the snip of information they give you to find HOW MUCH IS REPRESENTED BY EACH AREA BLOCK.
3) Divide all the bars into THE SAME SIZED AREA BLOCKS and so work out the number for each bar (using AREAS).

EXAMPLE:

The histogram below represents the age distribution of people arrested for slurping boiled sweets in public places in 1995. Given that there were 36 people in the 55 to 65 age range, find the number of people arrested in all the other age ranges.

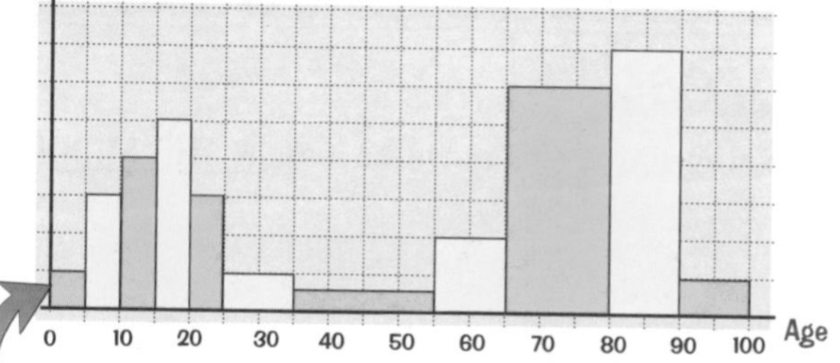

ANSWER: The 55-65 bar represents 36 people and contains 4 dotted squares, so each dotted square must represent 9 people.

The rest is easy. E.g. the 80-90 group has 14 dotted squares so that represents $14 \times 9 = 126$ people.

REMEMBER: ALWAYS COUNT AREA BLOCKS to find THE NUMBER IN EACH BAR.

The vertical axis is always called frequency density...

FREQUENCY DENSITY = FREQUENCY ÷ CLASS WIDTH

You don't need to worry too much about this. It says in the syllabus that you need to understand frequency density, so here it is. Learn the formula and you'll be fine.

Stem and Leaf Diagrams Use Numbers instead of Bars...

If you get one of these in the exam, you're laughing. It's the EASIEST THING IN THE WORLD.

1) Put the data in order.
2) Put it in groups and make a key.
3) Draw the diagram.

7, 11, 12, 13, 16, 17, 20, 23, 24, 24, 25, 26, 26, 29, 29, 31, 32, 34

This looks like it'll split nicely into tens: KEY: 2 | 3 = 23

Draw a line here.

Put the first digit of each group in a column.

Then put the second digits in rows like this.

```
0 | 7
1 | 1 2 3 6 7
2 | 0 3 4 4 5 6 6 9 9
3 | 1 2 4
```

This one means "26".

The Acid Test:

LEARN the THREE RULES for Histograms and the formula for frequency density. Turn over and write it all down.

1) Find the number of people in each of the age ranges for the histogram above.
2) Draw a stem and leaf diagram for this data: 3, 16, 14, 22, 7, 11, 26, 17, 12, 19, 20, 6, 13, 24, 26

Spread of Data

Scatter Graphs — correlation and the line of best fit

A scatter graph tells you how closely two things are related — the fancy word for this is CORRELATION. Good correlation means the two things are closely related to each other. Poor correlation means there is very little relationship. The LINE OF BEST FIT goes roughly through the middle of the scatter of points. (It doesn't have to go through any of the points exactly but it can.) If the line slopes up it's positive correlation, if it slopes down it's negative correlation. No correlation means there's no linear relationship.

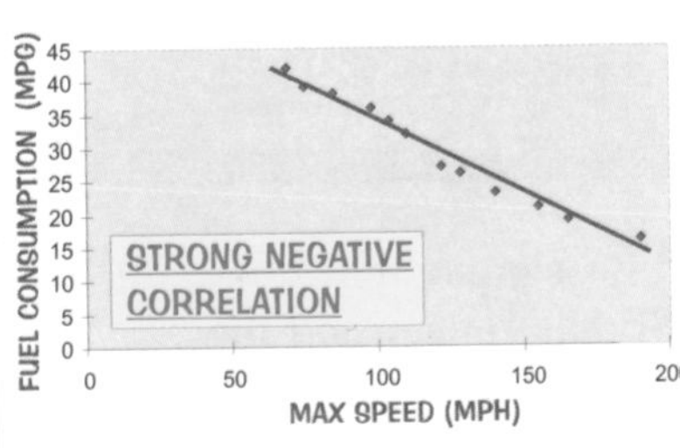

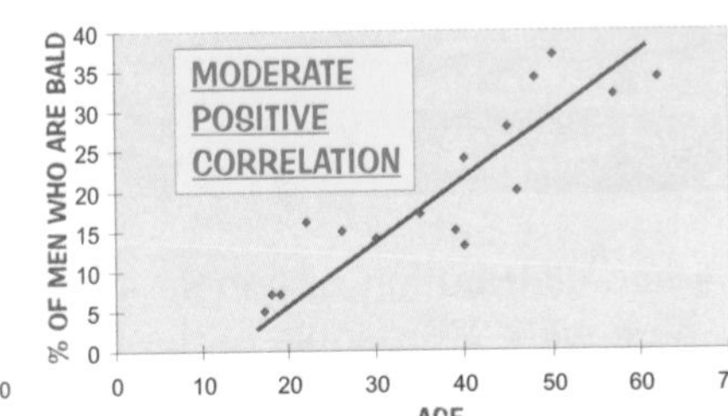

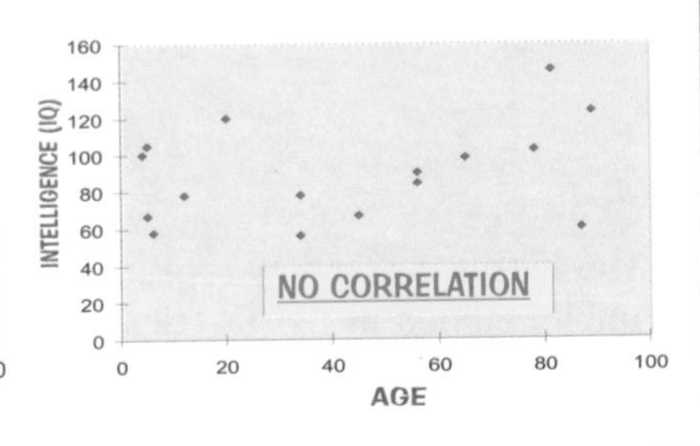

Shapes of Histograms and "Spread"

You can easily estimate the mean from the shape of a histogram — it's more or less IN THE MIDDLE.

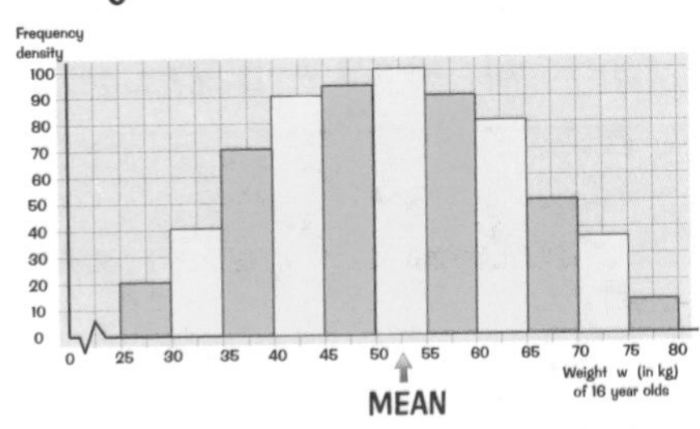

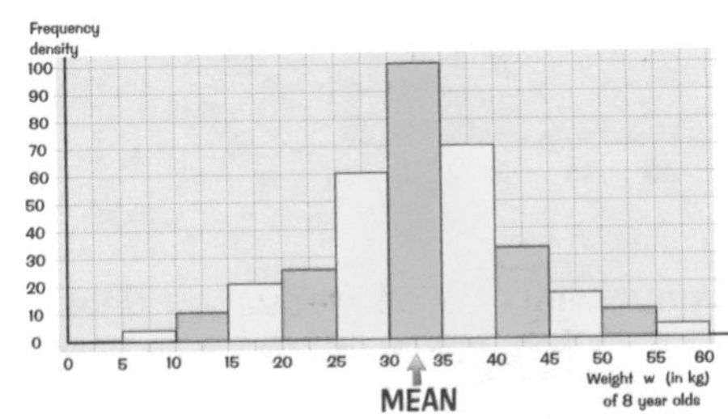

You must LEARN the significance of the shapes of these two histograms:

1) The first shows high dispersion (i.e. a large spread of results away from the mean).
   (i.e. the weights of a sample of 16 year olds will cover a very wide range)
2) The second shows a "tighter" distribution of results where most values are within a narrow range either side of the mean.
   (i.e the weights of a sample of 8 year olds will show very little variation)

Cumulative Freq. Curves and "Spread"

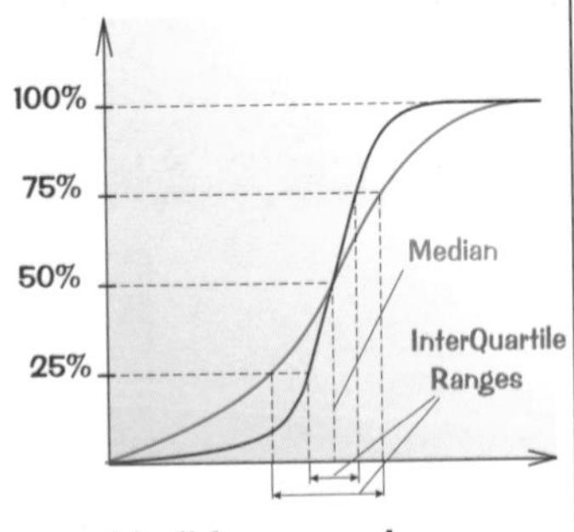

The shape of a CUMULATIVE FREQUENCY CURVE also tells us how spread out the data values are.

The blue curve shows a very tight distribution around the MEDIAN and this also means the inter-quartile range is small as shown.

The red curve shows a more widely spread set of data and therefore a larger interquartile range.

'Tight' distribution represents CONSISTENT results. E.g. the lifetimes of light bulbs would all be very close to the median, indicating a good product. The lifetimes of another product may show wide variation, which shows that the product is not as consistent. They often ask about this "shape significance" in Exams.

The Acid Test:

LEARN THIS PAGE. Then turn over and write down all the important details from memory.

1) Draw two contrasting histograms showing speeds of cyclists and motorists.
2) Sketch two cumulative frequency curves for heights of 5 yr olds and 13 yr olds.

Other Graphs and Charts

Two-Way Tables

Two-way tables are a bit like frequency tables, but they show two things instead of just one.

EXAMPLE:

"Use this table to work out how many
(a) right-handed people and
(b) left-handed women there were in this survey."

	Women	Men	TOTAL
Left-handed		27	63
Right-handed	164	173	
TOTAL	200	200	400

ANSWER:

a) 164 + 173 = 337 right-handed people (or you could have done 400 – 63 = 337).

b) 200 – 164 = 36 left-handed women (or you could have done 63 – 27 = 36). Easy.

Line Graphs and Frequency Polygons

A line graph is just a set of points joined with straight lines.

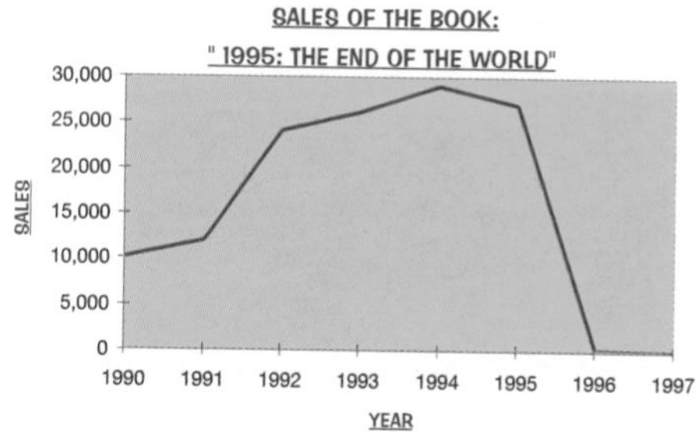

SALES OF THE BOOK:
" 1995: THE END OF THE WORLD"

(Axes: SALES 0 to 30,000; YEAR 1990 to 1997)

A frequency polygon looks similar and is used to show the information from a frequency table:

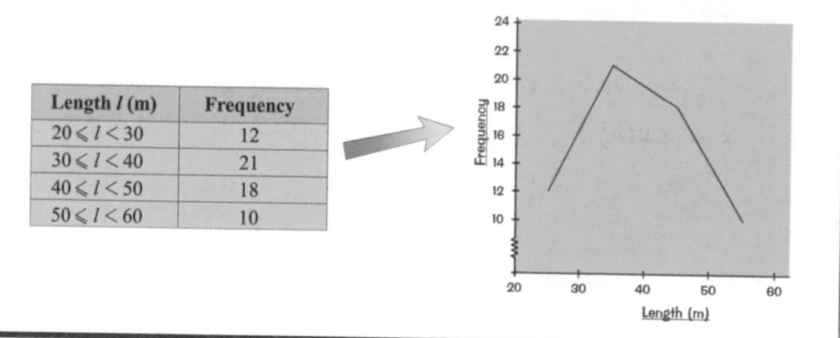

Length l (m)	Frequency
$20 \leqslant l < 30$	12
$30 \leqslant l < 40$	21
$40 \leqslant l < 50$	18
$50 \leqslant l < 60$	10

Pie Charts

Learn the Golden Rule for Pie Charts:

The TOTAL of Everything = 360°

Creature	Stick insects	Hamsters	Guinea Pigs	Rabbits	Ducks	Total
Number	12	20	17	15	26	90
		×4				×4
Angle		80°				360°

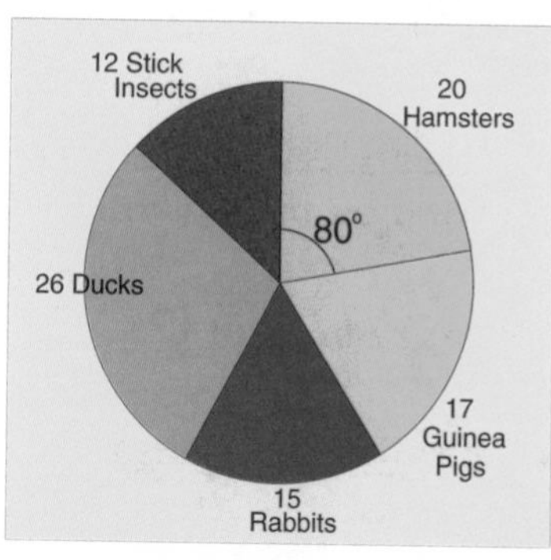

1) Add up all the numbers in each sector to get the TOTAL (← 90 for this one).
2) Then find the MULTIPLIER (or divider) that you need to turn your total into 360°:
   For 90 → 360 as above, the MULTIPLIER is 4.
3) Now MULTIPLY EVERY NUMBER BY 4 to get the angle for each sector.
   E.g. the angle for hamsters will be 20 × 4 = 80°.

The Acid Test:

LEARN ALL THE DETAILS of the three types of CHART.

1) Turn over the page and draw an example of each chart.
2) Work out the angles for all the other animals in the pie chart shown above.

Sampling Methods

This is all about doing surveys of 'populations' (not necessarily people) to find things out about them. Things start getting awkward when it's not possible to test the whole 'population', usually because there's just too many. In that case you have to take a sample, which means you somehow have to select a limited number of individuals so that they properly represent the whole 'population'.

There are FOUR DIFFERENT TYPES OF SAMPLING which you should know about:

RANDOM — this is where you just select individuals "at random". In practise it can be surprisingly difficult to make the selection truly random.

SYSTEMATIC — Start with a random selection and select every 10th or 100th one after that.

STRATIFIED — as in "strata" or "layers". E.g. to survey pupils in a school you would first pick a selection of the classes, and then pick students at random from those classes.

QUOTA — This is where you pick a sample which as far as possible reflects the whole population by having the same proportion of say males/females or adults/children etc.

Spotting Problems With Sampling Methods

In practise, the most important thing you should be able to do is to spot problems with sampling techniques, which means look for ways that the sample might not be a true reflection of the population. One mildly amusing way to practise, is to think up examples of bad sampling techniques:

1) A survey of motorists carried out in London concluded that 85% of the British people drive Black Cabs.

2) Two surveys carried out on the same street corner asked "Do you believe in God?" One found 90% of people didn't and the other found 90% of people did. The reason for the discrepancy? — one was carried out at 11pm Saturday night and the other at 10.15am Sunday morning.

3) A telephone survey carried out in the evening asked "What do you usually do after work or school?". It found that 80% of the population usually stay in and watch TV. A street survey conducted at the same time found that only 30% usually stay in and watch TV. Astonishing.

Other cases are less obvious:
In a telephone poll, 100 people were asked if they use the train regularly and 20% said yes. Does this mean 20% of the population regularly use the train?

ANSWER: Probably not. There are several things wrong with this sampling technique:

1) First and worst: the sample is far too small. At least 1000 would be more like it.
2) What about people who don't have their own phone?
3) What time of day was it done? When might regular train users be in or out?
4) Which part or parts of the country were telephoned?
5) If the results were to represent say the whole country then stratified or quota sampling would be essential.

The Acid Test: LEARN the names of the four sampling techniques together with their brief descriptions, and also the 5 points above.

1) A survey was done to investigate the average age of cars on Britain's roads by standing on a motorway bridge and noting the registration of the first 200 cars. Give three reasons why this is a poor sampling technique and suggest a better approach.

Time Series

Time Series — Measure the Same Thing over a Period of Time

A time series is what you get if you measure the same thing at a number of different times.

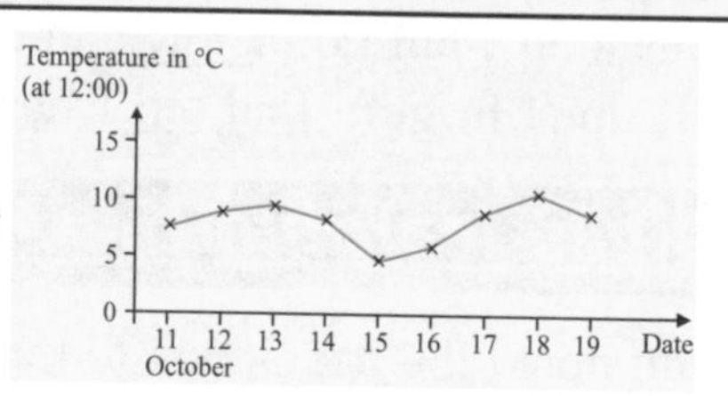

EXAMPLE: Measuring the temperature in your greenhouse at 12 o'clock each day gives you a time series — other examples might be profit figures, crime figures or rainfall.

THE RETAIL PRICE INDEX (RPI) IS A TIME SERIES: Every month, the prices of loads of items (same ones each month) — are combined to get an index number called the RPI, which is a kind of average. As goods get more expensive, this index number gets higher and higher. So when you see on TV that inflation this month is 2.5%, what it actually means is that the RPI is increasing at an annual rate of 2.5%.

Seasonality — The Same Basic Pattern

This is when there's a definite pattern that REPEATS ITSELF every so often. This is called SEASONALITY and the "so often" is called the PERIOD.

To find the PERIOD, measure PEAK TO PEAK (or trough to trough).

This series has a period of 12 months. There are a few irregularities, so the pattern isn't exactly the same every 12 months, but it's about right.

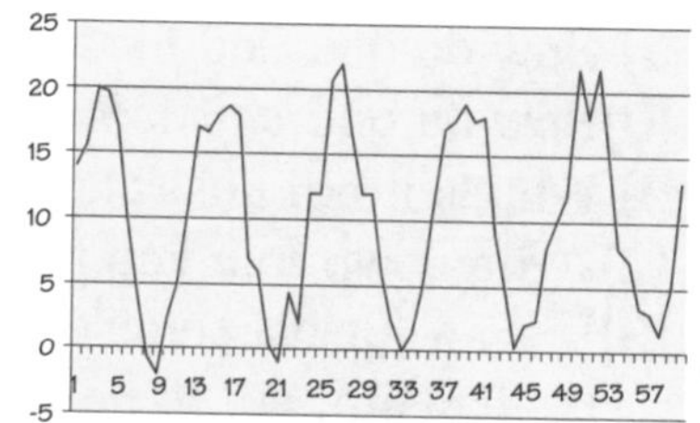

Trend — Ignoring the Wrinkles

This time series has lots of random fluctuations but there's a definite upwards trend.

The pink line is the trend line. It's straight, so this is a linear trend.

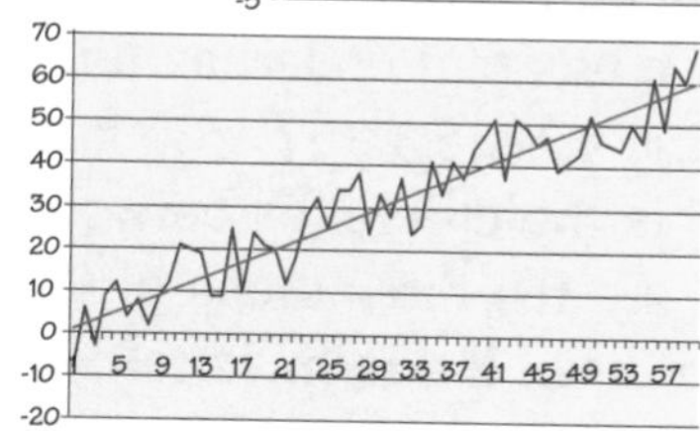

Moving Average — Smooths Out the Seasonality

It's easier to spot a trend if you can 'get rid of' the seasonality and some of the irregularities. One way to smooth the series is to use a moving average.

This is a time series that definitely looks periodic — but it's difficult to tell if there's a trend.

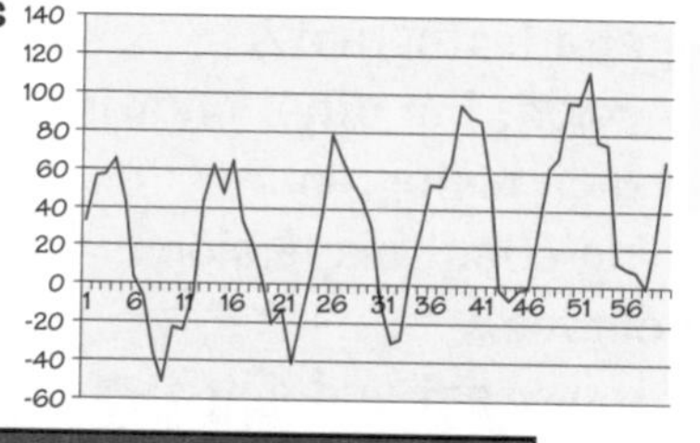

... but plot the moving average (in pink — must be pink — that's dead important)...

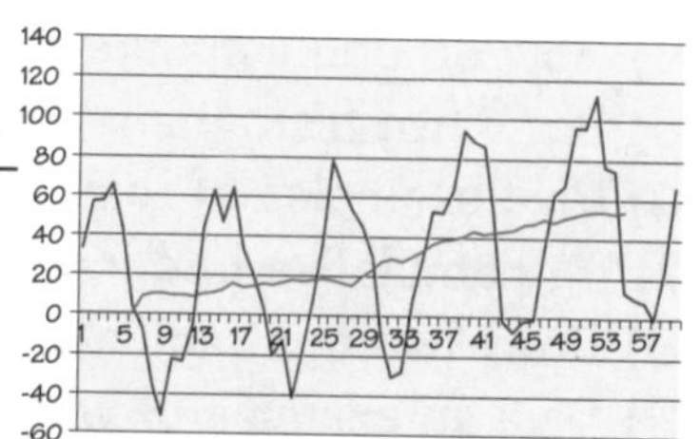

...and you can easily see the upward trend.

The period is 12, so you use 12 values for the moving average:

HOW TO FIND A MOVING AVERAGE:

Find the average of these 12 values... ...then of these... ...then of these, and so on.

month	1	2	3	4	5	6	7	8	9	10	11	12	13	14	...
temperature	38.00	42.30	59.00	32.30	25.00	2.00	-5.00	-51.30	-35.00	-45.30	-22.00	1.00	49.00	62.30	...

The Acid Test:

LEARN the words TIME SERIES, SEASONALITY, PERIOD, TREND, MOVING AVERAGE. Cover the page and write a description of each.

1) My town's rainfall is measured every month for 20 yrs and graphed. There's a rough pattern, which repeats itself every 4 months. a) What is the period of this time series? b) Describe how to calculate a moving average.

Unit Two Revision Summary

Here's the really fun page. The inevitable list of straight-down-the-middle questions to test how much you know. Remember, these questions will sort out (quicker than anything else can) exactly what you know and what you don't. And that's exactly what revision is all about, don't forget: find out what you DON'T know and then learn it until you do. Enjoy.

Keep learning these basic facts until you know them

1) Write down the definitions of mean, median, mode and range.
2) What is the Golden Rule for finding the above for a set of data?
3) How should you tackle all probability questions?
4) Write down four important facts about tree diagrams.
5) Draw a general tree diagram and put all the features on it.
6) There are four other important things to know about probability. What are they?
7) Write down eight important details about frequency tables.
8) What does $5 < x \leq 10$ mean?
9) How do you find the mid-interval values and what do you use them for?
10) How do you *estimate* the mean from a grouped frequency table?
11) Why is it not possible to find the exact value of the mean from a grouped frequency table?
12) Write down four key points about cumulative frequency tables.
13) Draw a typical cumulative frequency curve, and indicate on it exactly where the median etc. are to be found.
14) Draw a box plot underneath your cumulative frequency curve from question 13.
15) What is a histogram?
16) What is the difference between one of these and a regular bar chart?
17) What are the three steps of the method for tackling all histograms?
18) Write down the formula for frequency density.
19) What is a scatter graph?
20) What does a scatter graph illustrate? What is the fancy word for this?
21) Draw three examples to illustrate the three main types.
22) What is the meaning of dispersion?
23) Can you deduce anything about the dispersion of a set of data from the shape of the histogram?
24) How do you estimate the mean from looking at a histogram?
25) Draw two histograms, one showing high dispersion, the other showing tight distribution.
26) Give examples of real data that might match each histogram.
27) Do cumulative frequency curves tell us anything about dispersion?
28) Draw two contrasting cumulative frequency curves.
29) Give an example of what these curves might represent and say what the significant difference between the two things will be.
30) Which numerical figure represents dispersion on a cumulative frequency curve?
31) What is sampling all about? When is it needed?
32) Name the four main sampling methods, with a brief description of each.
33) List five common problems with conducting surveys.
34) Which one of these is NOT a time series?
    a) measuring the temperature in 20 different countries at 12:00 today, GMT,
    b) measuring the temperature in Britain at 12:00 every day for 100 days,
    c) the Retail Price Index.
35) How can you find out if a seasonal time series has an overall trend?

Types of Number

1) SQUARE NUMBERS:

(1x1)	(2x2)	(3x3)	(4x4)	(5x5)	(6x6)	(7x7)	(8x8)	(9x9)	(10x10)	(11x11)	(12x12)	(13x13)	(14x14)	(15x15)
1	4	9	16	25	36	49	64	81	100	121	144	169	196	225..

Differences: 3 5 7 9 11 13 15 17 19 21 23 25 27 29

Note that the DIFFERENCES between the square numbers are all the ODD numbers.

2) CUBE NUMBERS:

They're called CUBE NUMBERS because they're like the volumes of this pattern of cubes.

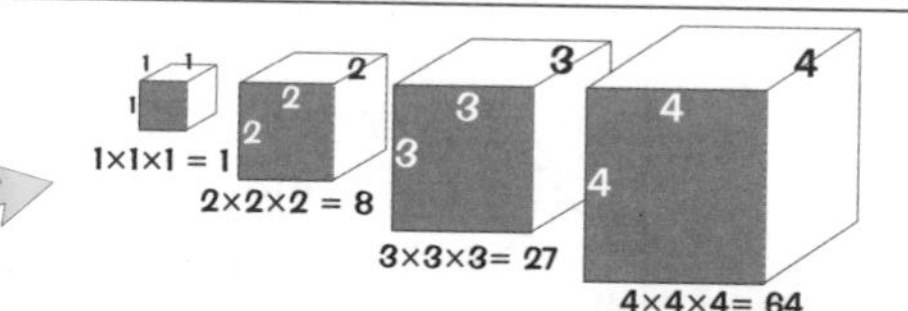

1 8 27 64 125 216 343 512 729 1000...

Admit it, you never knew maths could be this exciting did you.

3) TRIANGLE NUMBERS:

To remember the triangle numbers you have to picture in your mind this increasing pattern of triangles, where each new row has one more blob than the previous row.

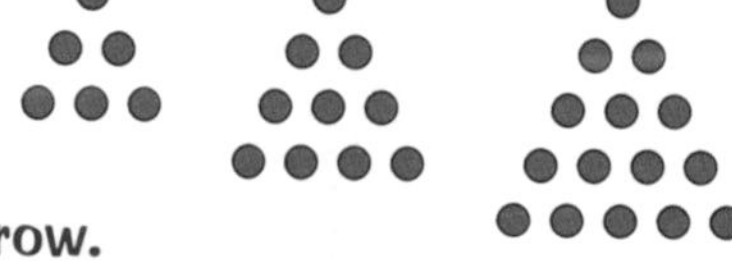

1 3 6 10 15 21 28 36 45 55...

Differences: 2 3 4 5 6 7 8 9 10

It's definitely worth learning this simple pattern of differences, as well as the formula for the nth term (see P.21) which is: nth term = $\frac{1}{2}n(n+1)$

4) POWERS:

Powers are "numbers multiplied by themselves so many times".

"Two to the power three" = $2^3 = 2 \times 2 \times 2 = 8$

Here's the first few POWERS OF 2:

2 4 8 16 32...

$2^1=2$ $2^2=4$ $2^3=8$ $2^4=16$ etc...

... and the first POWERS OF 10 (even easier):

10 100 1000 10 000 100 000...

$10^1=10$ $10^2=100$ $10^3=1000$ etc...

5) PRIME NUMBERS:

2 3 5 7 11 13 17 19 23 29 31 37 41 43...

Prime numbers only divide by themselves and 1 (Note that 1 is NOT a prime number).

Apart from 2 and 5, ALL PRIMES END IN 1, 3, 7, OR 9.

So POSSIBLE primes are: 71, 73, 77, 79, ... 101, 103, 107, 109 .... 241, 243, 247, 249 ... etc.

However, not all of those are primes, and working out which are and which aren't is a little bit tricky — see next page for all the details on how to find prime numbers.

The Acid Test:

LEARN THIS PAGE. Then turn it over and write down all the important details. See what you missed and try again.

1) Write down the next five numbers in each of the sequences on this page.
2) Write down an expression for the nth term for each of the first 3 sequences on this page.

Prime Numbers

1) Basically, PRIME Numbers don't divide by anything

And that's the best way to think of them. (Strictly, they divide by themselves and 1)
So prime numbers are all the numbers that don't come up in times tables:

2 3 5 7 11 13 17 19 23 29 31 37 ...

As you can see, they're an awkward-looking bunch (that's because they don't divide by anything!). For example:

The only numbers that multiply to give 7 are 1×7
The only numbers that multiply to give 31 are 1×31

In fact the only way to get ANY PRIME NUMBER is $1 \times$ ITSELF

2) They All End in 1, 3, 7 or 9

1) 1 is NOT a prime number.
2) The first four prime numbers are 2, 3, 5 and 7.
3) Prime numbers end in 1, 3, 7 or 9 (2 and 5 are the only exceptions to this rule).
4) But NOT ALL numbers ending in 1, 3, 7 or 9 are primes, as shown here: (Only the circled ones are primes)

(2)	(3)	(5)	(7)
(11)	(13)	(17)	(19)
21	(23)	27	(29)
(31)	33	(37)	39
(41)	(43)	(47)	49
51	(53)	57	(59)
(61)	63	(67)	69

3) How to Find Prime Numbers — a very simple method

For a chosen number to be a prime:

1) It must end in either 1, 3, 7, or 9.
2) It WON'T DIVIDE by any of the primes below the value of its own square root.

If something like this comes up on the non-calculator paper, you'll have to start by estimating the square root — see P. 44.

Example: "Decide whether or not 233 is a prime number."

1) Does it end in either 1, 3, 7 or 9? Yes
2) Find its square root: $\sqrt{233} = 15.264$
3) List all primes which are less than this square root: 2, 3, 5, 7, 11 and 13
4) Divide all of these primes into the number under test:
$233 \div 3 = 77.6667$ $233 \div 7 = 33.2857$
$233 \div 11 = 21.181818$ $233 \div 13 = 17.923077$
(it obviously won't divide by 2 or 5.)
5) Since none of these divide cleanly into 233 then it is a prime number. Easy Peasy

The Acid Test: LEARN the main points in ALL 3 SECTIONS above.

Now cover the page and write down everything you've just learned.
1) Write down the first 15 prime numbers (*without* looking them up).
2) Find all the prime numbers between a) 100 and 110 b) 200 and 210 c) 500 and 510

Multiples, Factors and Prime Factors

Multiples

The MULTIPLES of a number are simply its TIMES TABLE:

E.g. the multiples of 13 are 13 26 39 52 65 78 91 104 ...

Factors

The FACTORS of a number are all the numbers that DIVIDE INTO IT. There is a special way to find them:

EXAMPLE 1: "Find all the factors of 24"

Start off with 1 × the number itself, then try 2 × , then 3 × and so on, listing the pairs in rows like this. Try each one in turn and put a dash if it doesn't divide exactly. Eventually, when you get a number repeated, you stop.

Increasing by 1 each time

1 x 24
2 x 12
3 x 8
4 x 6
5 x -
6 x 4

So the FACTORS OF 24 are 1,2,3,4,6,8,12,24

This method guarantees you find them ALL. And don't forget 1 and 24!

EXAMPLE 2: "Find the factors of 64"

Check each one in turn, to see if it divides or not. Use your calculator if you are not totally confident.

1 × 64
2 × 32
3 × –
4 × 16
5 × –
6 × –
7 × –
8 × 8 — The 8 has repeated so stop here.

So the FACTORS of 64 are 1,2,4,8,16,32,64

Finding Prime Factors — The Factor Tree

Any number can be broken down into a string of prime numbers all multiplied together – this is called "Expressing it as a product of prime factors", and to be honest it's pretty tedious – but it's in the Exam.

The mildly entertaining "Factor Tree" method is best, where you start at the top and split your number off into factors as shown. Each time you get a prime you ring it and you finally end up with all the prime factors, which you can then arrange in order.

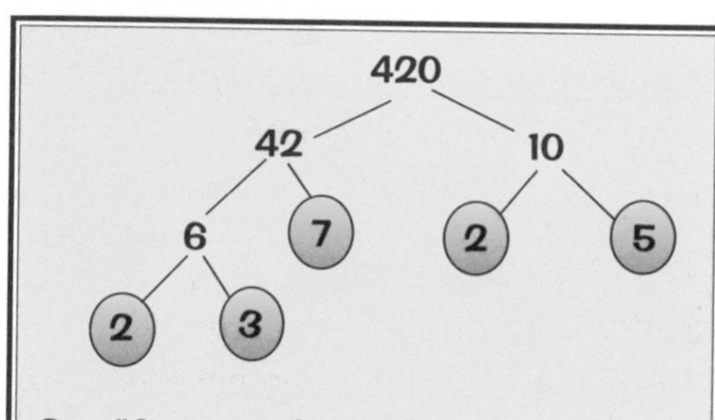

So, "As a product of prime factors", 420 = 2 × 2 × 3 × 5 × 7

The Acid Test:

LEARN what Multiples, Factors and Prime Factors are, AND HOW TO FIND THEM. Turn over and write it down.

Then try these without the notes:

1) List the first 10 multiples of 7 and of 9. What is their Lowest Common Multiple (LCM)?
2) List all the factors of 36 and 84. What is their Highest Common Factor (HCF)?
3) Express as a product of prime factors: a) 990 b) 160.

Fractions, Decimals and Percentages

The one word that could describe all these three is PROPORTION. Fractions, decimals and percentages are simply three different ways of expressing a proportion of something — and it's pretty important you should see them as closely related and completely interchangeable with each other. This table shows the really common conversions which you should know straight off without having to work them out:

Fraction	Decimal	Percentage
1/2	0.5	50%
1/4	0.25	25%
3/4	0.75	75%
1/3	0.33333... or $0.\dot{3}$	$33\frac{1}{3}$%
2/3	0.66666... or $0.\dot{6}$	$66\frac{2}{3}$%
1/10	0.1	10%
2/10	0.2	20%
X/10	0.X	X0%
1/5	0.2	20%
2/5	0.4	40%

⅓ and ⅔ have what're known as 'recurring' decimals — the same pattern of numbers carries on repeating itself forever. (Except here, the pattern's just a single 3 or a single 6. You could have, for instance: 0.143143143...)

The more of those conversions you learn, the better — but for those that you don't know, you must also learn how to convert between the three types. These are the methods:

Fraction → Decimal: Divide (use your calculator if you can) e.g. ½ is 1÷2 = 0.5

Decimal → Percentage: × by 100 e.g. 0.5 × 100 = 50%

Percentage → Decimal: ÷ by 100

Decimal → Fraction: The awkward one

Converting decimals to fractions is fairly easy to do when you have exact decimals. It's best illustrated by examples — you should be able to work out the rule...

$0.6 = {}^{6}/_{10}$ $0.3 = {}^{3}/_{10}$ $0.7 = {}^{7}/_{10}$ $0.X = {}^{X}/_{10}$, etc.

$0.12 = {}^{12}/_{100}$ $0.78 = {}^{78}/_{100}$ $0.45 = {}^{45}/_{100}$ $0.05 = {}^{5}/_{100}$, etc.

$0.345 = {}^{345}/_{1000}$ $0.908 = {}^{908}/_{1000}$ $0.024 = {}^{24}/_{1000}$ $0.XYZ = {}^{XYZ}/_{1000}$, etc.

These can then be cancelled down.

Scary-looking recurring decimals like 0.3333333 are actually just exact fractions in disguise. The method for converting them into fractions is covered on p.51 (have a look at this — you could be asked to do it in the unit 3 exam...)

The Acid Test:

LEARN the whole of the top table and the 4 conversion processes for FDP.

1) Turn the following decimals into fractions and reduce to their simplest form.
a) 0.6 b) 0.02 c) 0.77 d) 0.555 e) 5.6

Finding the nth term

"The nth term" is a formula with "n" in it which gives you every term in a sequence when you put different values for n in. There are two different types of sequence (for "nth term" questions) which have to be done in different ways:

Common Difference Type: "dn + (a – d)"

For any sequence such as 3, 7, 11, 15, where there is a COMMON DIFFERENCE:
4 4 4

you can always find "the nth term" using the formula: "nth term = dn + (a – d)"

1) "a" is simply the value of THE FIRST TERM in the sequence.
2) "d" is simply the value of THE COMMON DIFFERENCE between the terms.
3) To get the nth term, you just find the values of "a" and "d" from the sequence and stick them in the formula.
You don't replace n though — that wants to stay as n
4) — of course YOU HAVE TO LEARN THE FORMULA, but life is like that.

Example:

"Find the nth term of this sequence: 5, 8, 11, 14 ...."

ANSWER: 1) The formula is dn + (a-d)
2) The first term is 5, so a = 5. The common difference is 3 so d = 3
3) Putting these in the formula gives: 3n + (5–3)
so the nth term = 3n + 2

Changing Difference Type:

"a + (n–1)d + ½(n–1)(n–2)C"

If the number sequence is one where the difference between the terms is increasing or decreasing then it gets a whole lot more complicated (as you can see from the above formula which you'll have to learn!). This time there are THREE letters to fill in:

"a" is the FIRST TERM,
"d" is the FIRST DIFFERENCE (between the first two numbers),
"C" is the CHANGE BETWEEN ONE DIFFERENCE AND THE NEXT.

Example:

"Find the nth term of this sequence: 2, 5, 9, 14 ...."
3 4 5

ANSWER: 1) The formula is "a + (n–1)d + ½(n–1)(n–2)C"
2) The first term is 2, so a = 2. The first difference is 3 so d = 3
3) The differences increase by 1 each time so C = +1

Putting these in the formula gives: "2 + (n–1)3 + ½(n–1)(n–2)×1"
Which becomes: 2 + 3n – 3 + ½n^2 – 1½n + 1
Which simplifies to: ½n^2 + 1½n = ½n(n+3) so the nth term = ½n(n+3).

The Acid Test:

LEARN the definition of the nth term and the 4 steps for finding it, and LEARN THE FORMULA.

1) Find the nth term of the following sequences:
a) 4, 7, 10, 13.... b) 3, 8, 13, 18,.... c) 1, 3, 6, 10, 15,.... d) 3, 4, 7, 12,...

Calculator Buttons

These two pages are full of lovely calculator tricks to save you a lot of button-bashing.

The Fraction Button: a b/c

Use this as much as possible in the calculator paper. It's very easy, so make sure you know how to use it — you'll lose a lot of marks if you don't:

1) To enter $\frac{1}{4}$ press 1 a b/c 4
2) To enter $1\frac{3}{5}$ press 1 a b/c 3 a b/c 5
3) To work out $\frac{1}{5}\times\frac{3}{4}$ press 1 a b/c 5 X 3 a b/c 4 =
4) To reduce a fraction to its lowest terms enter it and then press =
   e.g. $\frac{9}{12}$, 9 a b/c 12 = [3⌟4] = $\frac{3}{4}$
5) To convert between mixed and top-heavy fractions press SHIFT a b/c
   e.g. $2\frac{3}{8}$ 2 a b/c 3 a b/c 8 = SHIFT a b/c which gives $\frac{19}{8}$

BODMAS and the BRACKETS BUTTONS (and)

One of the biggest problems people have with their calculators is not realising that the calculator always works things out IN A CERTAIN ORDER, which is summarised by the word BODMAS, which stands for:

Brackets, Other, Division, Multiplication, Addition, Subtraction

This is really important when you want to work out even a simple thing like $\frac{23+45}{64\times3}$

You can't just press 23 + 45 ÷ 64 × 3 = — it will be completely wrong.

The calculator will think you mean $23+\frac{45}{64}\times3$ because the calculator will do the division and multiplication BEFORE it does the addition.

The secret is to OVERRIDE the automatic BODMAS order of operations using the BRACKETS BUTTONS. Brackets are the ultimate priority in BODMAS, which means anything in brackets is worked out before anything else happens to it.

So all you have to do is:

1) Write a couple of pairs of brackets into the expression: $\frac{(23+45)}{(64\times3)}$
2) Then just type it as it's written:

(23 + 45) ÷ (64 × 3) =

You might think it's difficult to know where to put the brackets in.
It's not that difficult, you just put them in pairs around each group of numbers.
It's OK to have brackets within other brackets too, e.g. $(4 + (5\div2))$
As a rule, you can't cause trouble by putting too many brackets in,

SO LONG AS THEY ALWAYS GO IN PAIRS.

Calculator Buttons

The MEMORY BUTTONS (STO Store, RCL Recall)

These are really useful for keeping a number you've just calculated, so you can use it again shortly afterwards.

E.g. Find $\frac{840}{15+12\sin 40}$ — just work out the bottom line first and stick it in the memory.

So press 15 + 12 SIN 40 = and then STO M to keep the result of the bottom line in the memory. Then you simply press 840 ÷ RCL M =, and the answer is 36.98.

The memory buttons might work a bit differently on your calculator. Note, if your calculator has an "Ans" button, you can do the same thing as above using that instead — the Ans button gives you the result you got when you last pressed the "=" button.

Converting Time to Hrs, Mins and Secs with °' "

Here's a tricky detail that comes up when you're doing speed distance and time: converting an answer like 2.35 hours into hours and minutes. What it definitely ISN'T is 2 hours and 35 mins — remember your calculator does not work in hours and minutes unless you tell it to, as shown below. You'll need to practise with this button, but you'll be glad you did.

1) To ENTER a time in hours, mins and secs

E.g. 5hrs 34mins and 23 secs, press 5 °' " 34 °' " 23 °' " = to get 5° 34° 23.

2) Converting hours, min and secs to a decimal time:

Enter the number in hours, mins and secs as above.

Then just press °' " and it should convert it to a decimal like this 5.573055556.

(Though some older calculators will automatically convert it to decimal when you enter a time in hours, minutes and secs.)

3) To convert a decimal time (as you always get from a formula) into hrs, mins and secs:

E.g. To convert 2.35 hours into hrs, mins and secs.

Simply press 2.35 = to enter the decimal, then press SHIFT °' ".

The display should become 2° 21° 0 which means 2 hours, 21 mins (and 0 secs).

The Acid test:

LEARN your calculator buttons. Practise until you can answer all of these without having to refer back:

1) Convert these into top-heavy fractions: a) 2 ¾ b) 16 ½ c) 8 ¼

2) Explain what STO and RCL do and give an example of using them.

3) Write down what buttons you would press to work this out in one go: $\frac{23.3+35.8}{36\times 26.5}$

4) a) Convert 4.57 hrs into hrs and mins.

b) Convert 5hrs 32mins and 23secs into decimal hrs.

Conversion Factors

Conversion Factors are a mighty powerful tool for dealing with a wide variety of questions. And what's more the method is real easy. Learn it now. It's ace.

1) Find the Conversion Factor (always easy)
2) Multiply by it AND divide by it
3) Choose the common sense answer

Three Important Examples

1) "Convert 2.55 hours into minutes." — (N.B. This is NOT 2hrs 55mins)

1) Conversion factor = 60 — (simply because 1 hour = 60 mins)
2) 2.55 hrs × 60 = 153mins (makes sense)
   2.55 hrs ÷ 60 = 0.0425 mins (ridiculous answer!)
3) So plainly the answer is that 2.55hrs = 153 mins

2) "If £1 = 1.7 US Dollars, how much is 63 US Dollars in £s?"

1) Obviously, Conversion Factor = 1.7 (The "exchange rate")
2) 63 × 1.7 = £107.10
   63 ÷ 1.7 = £37.06
3) Not quite so obvious this time, but since 1.7 US Dollars = £1, you're clearly going to have less pounds than you had Dollars (roughly half). So the answer has to be less than 63, which means it must be **£37.06**

3) "A map has a scale of 1:20,000. How big in real life is a distance of 3cm on the map?"

1) Conversion Factor = 20 000
2) 3cm × 20 000 = 60 000cm (looks OK)
   3cm ÷ 20 000 = 0.00015cm (not good)
3) So 60,000cm is the answer.
   How do we convert to metres? →

To Convert 60,000cm to m:

1) C.F. = 100 (cm ⟷ m)
2) 60,000 × 100 = 6,000,000m (hmm)
   60,000 ÷ 100 = 600m (more like it)
3) So answer = 600m

The Acid Test:

LEARN the 3 steps of the Conversion Factor method. Then turn over and write them down.

1) Convert 2.3 km into m. 2) Which is more, £34 or €45 ? (Exchange rate: £1 = €1.4)
3) A map is drawn to a scale of 2 cm = 5 km. A road is 8 km long.
How many cm will this be on the map? (Hint, C.F. = 5÷2, i.e. 1 cm = 2.5 km)

Metric and Imperial Units

Make sure you learn all these easy facts:

Metric Units

1) Length mm, cm, m, km
2) Area mm^2, cm^2, m^2, km^2,
3) Volume mm^3, cm^3, m^3, litres, ml
4) Weight g, kg, tonnes
5) Speed km/h, m/s

MEMORISE THESE KEY FACTS:

1cm = 10mm
1m = 100cm
1km = 1000m
1kg = 1000g
1 tonne = 1000kg
1 litre = 1000ml
1 litre = $1000cm^3$
1 cm^3 = 1 ml

Imperial Units

1) Length Inches, feet, yards, miles
2) Area Square inches, square feet, square yards, square miles
3) Volume Cubic inches, cubic feet, gallons, pints
4) Weight Ounces, pounds, stones, tons
5) Speed mph

LEARN THESE TOO!

1 Foot = 12 Inches
1 Yard = 3 Feet
1 Gallon = 8 Pints
1 Stone = 14 Pounds (lbs)
1 Pound = 16 Ounces (Oz)

Metric-Imperial Conversions

YOU NEED TO LEARN THESE — they don't promise to give you these in the Exam and if they're feeling mean (as they often are), they won't.

APPROXIMATE CONVERSIONS

1 kg = 2¼ lbs
1m = 1 yard (+ 10%)
1 litre = 1¾ pints
1 inch = 2.5 cm
1 gallon = 4.5 litres
1 foot = 30cm
1 metric tonne = 1 imperial ton
1 mile = 1.6km
or 5 miles = 8 km

Using Metric-Imperial Conversion Factors

1) Convert 45mm into cm. CF = 10, so × and ÷ by 10, to get 450cm or 4.5cm. (Sensible)
2) Convert 37 inches into cm. CF = 2.5, so × and ÷ by 2.5, to get 14.8cm or 92.5cm.
3) Convert 5.45 litres into pints CF = 1¾, so × and ÷ by 1.75, to get 3.11 or 9.54 pints.

The Acid Test:

LEARN the 21 Conversion factors in the shaded boxes above. Then turn over and write them down.

1) How many litres is 3½ gallons?
2) Roughly how many yards is 200m?
3) A rod is 46 inches long. What is this in cm?
4) Petrol costs £4.92 per gallon. What should it cost per litre?
5) A car travels at 65 mph. What is its speed in km/h?

Rounding Numbers

There are two different ways of specifying where a number should be rounded off. They are: "Decimal Places" and "Significant Figures". Whichever way is used, the basic method is always the same and is shown below:

The Basic Method Has Three Steps

1) Identify the position of the LAST DIGIT.

2) Then look at the next digit to the RIGHT – called the DECIDER.

3) If the DECIDER is 5 or more, then ROUND-UP the LAST DIGIT.
   If the DECIDER is 4 or less, then leave the LAST DIGIT as it is.

EXAMPLE: "What is 7.45839 to 2 Decimal Places?"

7.45839 = 7.46

LAST DIGIT to be written (2nd decimal place because we're rounding to 2 D P)

DECIDER

The LAST DIGIT rounds UP because the DECIDER is 5 or more

Decimal Places (D.P)

This is pretty easy:

1) To round off to, say, 4 decimal places, the LAST DIGIT will be the 4th one after the decimal point.
2) There must be no more digits after the LAST DIGIT (not even zeros).

DECIMAL PLACES EXAMPLES

Original number: 45.319461

Rounded to 5 decimal places (5 d p)	45.31946	(DECIDER was 1, so don't round up)
Rounded to 4 decimal places (4 d p)	45.3195	(DECIDER was 6, so do round up)
Rounded to 3 decimal places (3 d p)	45.319	(DECIDER was 4, so don't round up)
Rounded to 2 decimal places (2 d p)	45.32	(DECIDER was 9, so do round up)

The Acid Test:

LEARN the 3 Steps of the Basic Method and the 2 Extra Points for Decimal Places.

Now turn over and write down what you've learned. Then try again till you know it.

1) Round 3.5743 to 2 decimal places
2) Give 0.0481 to 2 decimal places
3) Express 12.9096 to 3 DP
4) Express 3546.054 to 1 d.p.

Rounding Numbers

Significant Figures (Sig. Fig.)

The method for sig. fig. is identical to that for DP except that finding the position of the LAST DIGIT is more difficult — it wouldn't be so bad, but for the ZEROS ...

1) The 1st significant figure of any number is simply THE FIRST DIGIT WHICH ISN'T A ZERO.

2) The 2nd, 3rd, 4th, etc. significant figures follow on immediately after the 1st, REGARDLESS OF BEING ZEROS OR NOT ZEROS.

e.g **0.002309** — SIG FIGS: 1st (2), 2nd (3), 3rd (0), 4th (9)

2.03070 — SIG FIGS: 1st (2), 2nd (0), 3rd (3), 4th (0)

(If we're rounding to say, 3 sig. fig. then the LAST DIGIT is simply the 3rd sig. fig.)

3) After Rounding Off the LAST DIGIT, end ZEROS must be filled in up to, BUT NOT BEYOND, the decimal point.

No extra zeros must ever be put in after the decimal point.

Examples	to 4 SF	to 3 SF	to 2 SF	to 1 SF
1) 54.7651	54.77	54.8	55	50
2) 17.0067	17.01	17.0	17	20
3) 0.0045902	0.004590	0.00459	0.0046	0.005
4) 30895.4	30900	30900	31000	30000

POSSIBLE ERROR OF HALF A UNIT WHEN ROUNDING

Whenever a measurement is rounded off to a given UNIT the actual measurement can be anything up to HALF A UNIT bigger or smaller.

Examples:

1) A room is given as being "9m long to the nearest METRE" — its actual length could be anything from 8.5m to 9.5m — i.e. HALF A METRE either side of 9m.

2) If it was given as "9.4m, to the nearest 0.2m", then it could be anything from 9.3m to 9.5m — i.e. 0.1m either side of 9.4m.

3) "A school has 460 pupils to 2 Sig Fig" (i.e. to the nearest 10) — the actual figure could be anything from 455 to 464. — (Why isn't it 465?)

The Acid Test:

LEARN the whole of this page, then turn over and write down everything you've learned. It's all good clean fun.

1) Round these to 2 D.P: a) 3.408 b) 1.051 c) 0.068 d) 3.596

2) Round these to 3 S.F, and for each one say which of the 3 rules about ZEROS applies: a) 567.78 b) 23445 c) 0.04563 d) 0.90876

3) A car is described as 17 feet long to the nearest foot. What is the longest and shortest it could be, in feet and inches? (e.g. 14 feet 4 inches)

Basic Algebra

Negative numbers crop up everywhere so you need to learn this rule for dealing with them:

+	+	makes	+
+	–	makes	–
–	+	makes	–
–	–	makes	+

Only to be used when:

1) Multiplying or dividing:

e.g. $-2 \times 3 = -6$, $-8 \div -2 = +4$ $-4p \times -2 = +8p$

2) Two signs are together:

e.g. $5 - -4 = 5+4 = 9$ $4 + -6 - -7 = 4 - 6 + 7 = 5$

Letters Multiplied Together

Watch out for these combinations of letters in algebra that regularly catch people out:

1) abc means $a \times b \times c$. The ×'s are often left out to make it clearer.

2) gn^2 means $g \times n \times n$. Note that only the n is squared, not the g as well, e.g. πr^2

3) $(gn)^2$ means $g \times g \times n \times n$. The brackets mean that BOTH letters are squared.

4) $p(q - r)^3$ means $p \times (q - r) \times (q - r) \times (q - r)$. Only the brackets get cubed.

5) -3^2 is a bit ambiguous. It should either be written $(-3)^2 = 9$, or $-(3^2) = -9$

D.O.T.S. — The Difference Of Two Squares:

$$a^2 - b^2 = (a + b)(a - b)$$

The "difference of two squares" (D.O.T.S. for short) is where you have "one thing squared" take away "another thing squared". Too many people have more trouble than they should with this, probably because they don't make enough effort to learn it as a separate item in its own right. Best learn it now, eh, before it's too late.

1) Factorise $9P^2 - 16Q^2$. Answer: $9P^2 - 16Q^2 = (3P + 4Q)(3P - 4Q)$

2) Factorise $1 - T^4$. Answer: $1 - T^4 = (1 + T^2)(1 - T^2)$

3) Factorise $3K^2 - 75H^2$. Answer: $3K^2 - 75H^2 = 3(K^2 - 25H^2) = 3(K + 5H)(K - 5H)$

The Acid Test: LEARN everything on this page.

Now turn over and write it all down. Then do these without a calculator:

1) a) -4×-3 b) $-4 + -5 + 3$ c) $(3x + -2x - 4x) \div (2+-5)$ d) $120 \div -40$

2) If $m=2$ and $n=-3$ work out: a) mn^2 b) $(mn)^3$ c) $m(4+n)^2$ d) n^3 e) $3m^2n^3 + 2mn$

3) Factorise: a) $x^2 - 16y^2$ b) $49 - 81p^2q^2$ c) $12y^2x^6 - 48k^4m^8$

Basic Algebra

1) Terms

Before you can do anything else, you must understand what a term is:

1) **A TERM IS A COLLECTION OF NUMBERS, LETTERS AND BRACKETS, ALL MULTIPLIED/DIVIDED TOGETHER.**

2) Terms are separated by + and – signs E.g. $4x^2 - 3py - 5 + 3p$

3) Terms always have a + or – attached to the front of them

4) E.g. $4xy$ $+ 5x^2$ $- 2y$ $+ 6y^2$ $+ 4$

Invisible + sign; "xy" term; "x^2" term; "y" term; "y^2" term; "number" term

2) Simplifying — or "Collecting Like Terms"

EXAMPLE: Simplify $2x - 4 + 5x + 6$

Invisible + sign

$2x$ -4 $+5x$ $+6$ = $+2x$ $+5x$ -4 $+6$

x-terms, number terms = $7x$ $+2$ = $7x + 2$

1) Put bubbles round each term — be sure you capture the +/– sign in front of each.
2) Then you can move the bubbles into the best order so that like terms are together.
3) "Like terms" have exactly the same combination of letters, e.g. x-terms or xy-terms.
4) Combine like terms using the number line (not the other rule for negative numbers).

3) Multiplying out Brackets

1) The thing outside the brackets multiplies each separate term inside the brackets.
2) When letters are multiplied together, they are just written next to each other, pq.
3) Remember, $R \times R = R^2$, and TY^2 means $T\times Y\times Y$, whilst $(TY)^2$ means $T\times T\times Y\times Y$.
4) Remember a minus outside the bracket REVERSES ALL THE SIGNS when you multiply.

1) $3(2x + 5) = 6x + 15$ 2) $4p(3r - 2t) = 12pr - 8pt$

3) $-4(3p^2 - 7q^3) = -12p^2 + 28q^3$ (note both signs have been reversed — Rule 4)

5) DOUBLE BRACKETS — you get 4 terms, and usually 2 of them combine to leave 3 terms.

$$(2P - 4)(3P + 1) = (2P\times3P) + (2P\times1) + (-4\times3P) + (-4\times1)$$
$$= 6P^2 + 2P - 12P - 4$$
$$= 6P^2 - 10P - 4$$

(these 2 combine together)

6) SQUARED BRACKETS — Always write these out as TWO BRACKETS:

E.g. $(3d + 5)^2$ should be written out as $(3d + 5)(3d + 5)$ and then work them out as above.
YOU SHOULD ALWAYS GET FOUR TERMS from a pair of brackets.
The usual WRONG ANSWER is $(3d + 5)^2 = 9d^2 + 25$ (eeek)
It should be: $(3d + 5)^2 = (3d + 5)(3d + 5) = 9d^2 + 15d + 15d + 25 = 9d^2 + 30d + 25$

Basic Algebra

4) Factorising — putting brackets in

This is the exact reverse of multiplying-out brackets. Here's the method to follow:

1) Take out the biggest number that goes into all the terms.
2) Take each letter in turn and take out the highest power (e.g. x, x^2 etc) that will go into EVERY term.
3) Open the brackets and fill in all the bits needed to reproduce each term.

EXAMPLE: "Factorise $15x^4y + 20x^2y^3z - 35x^3yz^2$"

Answer: $5x^2y(3x^2 + 4y^2z - 7xz^2)$

Biggest number that'll divide into 15, 20 and 35.

Highest powers of x and y that will go into all three terms.

z was not in ALL terms so it can't come out as a common factor.

REMEMBER:
1) The bits taken out and put at the front are the common factors.
2) The bits inside the brackets are what's needed to get back to the original terms if you multiplied the brackets out again.

5) Algebraic Fractions

The basic rules are exactly the same as for ordinary fractions.

1) Multiplying (easy)

Multiply top and bottom separately and cancel if possible:

e.g. $\frac{st}{10w^3} \times \frac{35s^2tw}{6} = \frac{35s^3t^2w}{60w^3} = \frac{7s^3t^2}{12w^2}$

2) Dividing (easy)

Turn the second one upside down, then multiply and cancel if possible:

e.g. $\frac{12}{p+4} \div \frac{4(p-3)}{3(p+4)} = \frac{\not{12}\,3}{\not{p+4}} \times \frac{3(\not{p+4})}{\not{4}(p-3)} = \frac{9}{p-3}$

3) Adding/subtracting (not so easy)

Always get a common denominator i.e. same bottom line (by cross-multiplying) and then ADD TOP LINES ONLY:

e.g. $\frac{t-2p}{3t-p} - \frac{1}{3} = \frac{3(t-2p)}{3(3t-p)} - \frac{1(3t-p)}{3(3t-p)} = \frac{3t-6p-3t+p}{3(3t-p)} = \frac{-5p}{3(3t-p)}$

The Acid Test:

LEARN THE DETAILS of all 5 sections on these two pages. Then turn over and write down what you've learned.

1) Simplify: $5x + 3y - 4 - 2y - x$
2) Expand $2pq(3p - 4q^2)$
3) Expand $(2g+5)(4g-2)$
4) Factorise $14x^2y^3 + 21xy^2 - 35x^3y^4$
5) Simplify $\frac{5abc^3}{18de} \div \frac{15abd^2}{9ce}$
6) Simplify $\frac{3}{5} + \frac{5g}{3g-4}$

Powers and Roots

Powers are a very useful shorthand: $2 \times 2 \times 2 \times 2 \times 2 \times 2 \times 2 = 2^7$ ("two to the power 7")

That bit is easy to remember. Unfortunately, there are TEN SPECIAL RULES for Powers that are not tremendously exciting, but you do need to know them for the Exam:

The Seven Easy Rules:

The first two only work for powers of the same number.

1) When MULTIPLYING, you ADD THE POWERS. e.g. $3^4 \times 3^6 = 3^{6+4} = 3^{10}$
2) When DIVIDING, you SUBTRACT THE POWERS. e.g. $5^4 \div 5^2 = 5^{4-2} = 5^2$
3) When RAISING one power to another, you MULTIPLY THEM. e.g. $(3^2)^4 = 3^{2\times4} = 3^8$
4) $X^1 = X$, ANYTHING to the POWER 1 is just ITSELF. e.g. $3^1 = 3$, $6 \times 6^3 = 6^4$
5) $X^0 = 1$, ANYTHING to the POWER 0 is just ONE. e.g. $5^0 = 1$ $67^0 = 1$
6) $1^x = 1$, 1 TO ANY POWER is STILL JUST 1. e.g. $1^{23} = 1$ $1^{89} = 1$ $1^2 = 1$
7) FRACTIONS — Apply Power to both TOP and BOTTOM. e.g. $(1\tfrac{3}{5})^3 = \left(\tfrac{8}{5}\right)^3 = \tfrac{8^3}{5^3} = \tfrac{512}{125}$

The Three Tricky Rules:

8) NEGATIVE POWERS —TURN IT UPSIDE-DOWN

People do have quite a bit of difficulty remembering this.
Whenever you see a negative power you're supposed to immediately think:
"Aha, that means turn it the other way up and make the power positive"
Like this: e.g. $7^{-2} = \frac{1}{7^2} = \frac{1}{49}$ $\left(\frac{3}{5}\right)^{-2} = \left(\frac{5}{3}\right)^{+2} = \frac{5^2}{3^2} = \frac{25}{9}$

9) FRACTIONAL POWERS

The Power ½ means Square Root,
The Power $\frac{1}{3}$ means Cube Root,
The Power ¼ means Fourth Root etc.

e.g. $25^{\frac{1}{2}} = \sqrt{25} = 5$
$64^{\frac{1}{3}} = \sqrt[3]{64} = 4$
$81^{\frac{1}{4}} = \sqrt[4]{81} = 3$ etc.

The one to really watch is when you get a negative fraction like $49^{-\frac{1}{2}}$ — people get mixed up and think that the minus is the square root, and forget to turn it upside down as well.

10) TWO-STAGE FRACTIONAL POWERS

They really like putting these in Exam questions so learn the method:
With fractional powers like $64^{\frac{5}{6}}$ always split the fraction into a root and a power, and do them in that order: root first, then power: $(64)^{\frac{1}{6}\times5} = \left(64^{\frac{1}{6}}\right)^5 = (2)^5 = 32$

Square Roots can be Positive or Negative

Whenever you take the square root of a number, the answer can be positive or negative...

E.g. $x^2 = 4$ gives $x = \pm\sqrt{4} = +2$ or -2

You always get a +ve and −ve version of the same number (your calculator only gives the +ve answer).

The reason for it becomes clear when you work backwards by squaring the answers:

$2^2 = 2 \times 2 = 4$ but also $(-2)^2 = (-2) \times (-2) = 4$

The Acid Test:

LEARN ALL TEN Exciting Rules on this page. Then TURN OVER and write them all down with examples. Keep trying till you can.

1) Simplify: a) $3^2 \times 3^6$ b) $4^3 / 4^2$ c) $(8^3)^4$ d) $(3^2 \times 3^3 \times 1^6)/3^5$ e) $7^3 \times 7 \times 7^2$
2) Evaluate a) $(\frac{1}{4})^{-3}$ b) 25^{-2} c) $25^{-\frac{1}{2}}$ d) $\left(\frac{27}{216}\right)^{-\frac{1}{3}}$ e) $625^{\frac{3}{4}}$ f) $125^{-\frac{2}{3}}$
3) Use your calculator to find: a) 5.2^{24} b) $40^{\frac{3}{4}}$ c) $\sqrt[5]{200}$

Standard Index Form

Standard form (or "standard index form") is only really useful for writing VERY BIG or VERY SMALL numbers in a more convenient way, e.g.

56,000,000,000 would be 5.6×10^{10} in standard form.
0.000 000 003 45 would be 3.45×10^{-9} in standard form.

but ANY NUMBER can be written in standard form and you need to know how to do it:

What it Actually is:

A number written in standard form must ALWAYS be in EXACTLY this form:

$$A \times 10^{n}$$

This number must always be BETWEEN 1 AND 10.
(The fancy way of saying this is: $1 \leq A < 10$ — they sometimes write that in Exam questions — don't let it put you off, just remember what it means).

This number is just the NUMBER OF PLACES the Decimal Point moves.

Learn The Three Rules:

1) The front number must always be BETWEEN 1 AND 10.

2) The power of 10, n, is purely: HOW FAR THE D.P. MOVES.

3) n is +ve for BIG numbers, n is –ve for SMALL numbers.

(This is much better than rules based on which way the D.P. moves.)

Two Very Simple Examples:

1) "Express 35 600 in standard form."

METHOD:
1) Move the D.P. until 35 600 becomes 3.56 ("$1 \leq A < 10$")
2) The D.P. has moved 4 places so n=4, giving: 10^4
3) 35600 is a BIG number so n is +4, not –4

ANSWER:
3.5 6 0 0.
$= 3.56 \times 10^4$

2) "Express 0.000623 in standard form."

METHOD:
1) The D.P. must move 4 places to give 6.23 ("$1 \leq A < 10$").
2) So the power of 10 is 4
3) Since 0.000623 is a SMALL NUMBER it must be 10^{-4} not 10^{+4}.

ANSWER:
0.000623
$= 6.23 \times 10^{-4}$

Standard Index Form

Four Very Important Examples

1) The Calculator's Scientific Mode

This mode gives all numbers in standard form to a specified number of sig fig.
A little SCI will be displayed somewhere when you're in this mode.
To get into this mode, press MODE and select SCI from one of the menus you get.
(On other calculators look for a button with "SCI" written above it as the 2nd or 3rd function.)
It'll ask you for the number of sig figs to display, something like this: SCI 0-9?
So if you choose 4, all numbers and answers will be displayed to 4 sig fig.

EXAMPLE: 565 ÷ 3 would give 188.3333333 in normal mode,
...or 1.883^02 in 4 sig fig mode.

2) What is 146.3 million in standard form?

The two favourite wrong answers for this are:

1) " 146.3×10^6 " which is kind of right but it's not in STANDARD FORM because 146.3, is not between 1 and 10 (i.e. $1 \leq A < 10$ has not been done)
2) " 1.463×10^6 " This one is in standard form but it's not big enough.

This is a very typical Exam question, which too many people get wrong.
Just take your time and do it IN TWO STAGES like this:

ANSWER: 146.3 million = 146,300,000 = 1.463×10^8

3) Remember, 10^5 means 1×10^5

So to enter 10^5 into the calculator you must remember it's actually 1×10^5 and press 1 EXP 5

EXAMPLE: "A nanometre is 10^{-9} m. How many nanometres are there in 0.35m? "
ANSWER: $0.35 \div (1 \times 10^{-9})$, so press 0.35 ÷ 1 EXP (–) 9 = $= 3.5 \times 10^8$.

4) The "Googol" is 10^{100} and is a pest

It's a pest because it goes off the scale of your calculator, so you have to do it "by hand" — which means they like it for Exam questions. So make sure you LEARN this example:
"Express 56 Googols in standard form."

ANS: 56 googols is $56 \times 10^{100} = 5.6 \times 10 \times 10^{100} = 5.6 \times 10^{101}$.
Note: you split the 56 into 5.6×10 and then COMBINE THE POWERS OF 10

The Acid Test:

LEARN the Three Rules and The Four Important Examples, then turn over and write them down.

1) Express 0.854 million and 0.00018 in standard index form.
2) Express 4.56×10^{-3} and 2.7×10^5 as ordinary numbers.
3) a) Work out $3.2 \times 10^7 \div 1.6 \times 10^{-4}$. b) How many nanometres are there in 10^{-1} m?
4) Write down 650 Googols in Standard Form.

Areas, Solids and Nets

LEARN THESE FORMULAS for calculating the areas of rectangles and triangles.

1) Rectangle

Area of rectangle = length × width

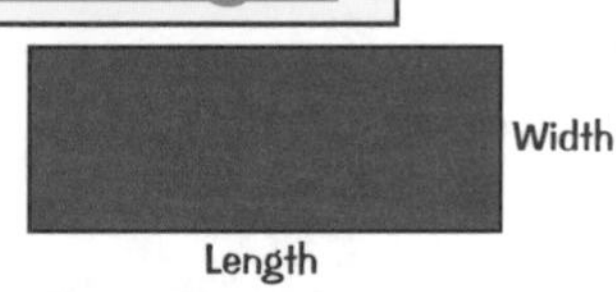

$$A = l \times w$$

2) Triangle

Area of triangle = ½ × base × vertical height

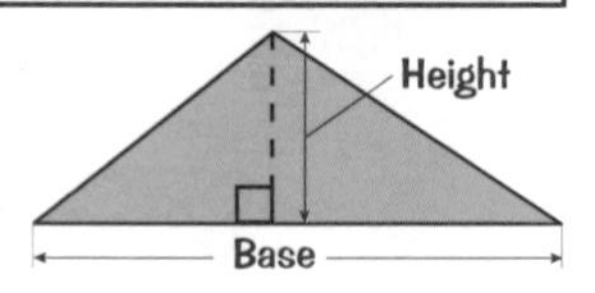

$$A = \frac{1}{2} \times b \times h_v$$

Note that the height must always be the vertical height, not the sloping height.

Surface Area and Nets

1) SURFACE AREA only applies to solid 3D objects, and it's simply the total area of all the outer surfaces added together.
2) There isn't usually a simple formula for surface area — you have to work out each side in turn and then add them all together.
3) A NET is just a solid shape folded out flat.
4) So obviously: SURFACE AREA OF SOLID = AREA OF NET.

The surface area for each of the shapes below can be found simply by using the formulas for rectangles and triangles, and adding the sides.

1) Triangular Prism

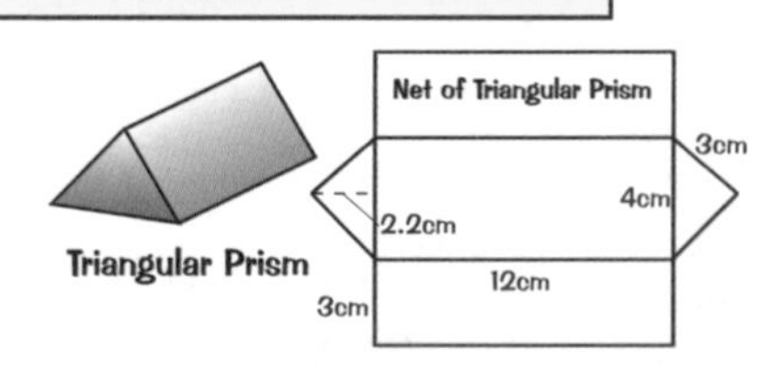

2) Cube

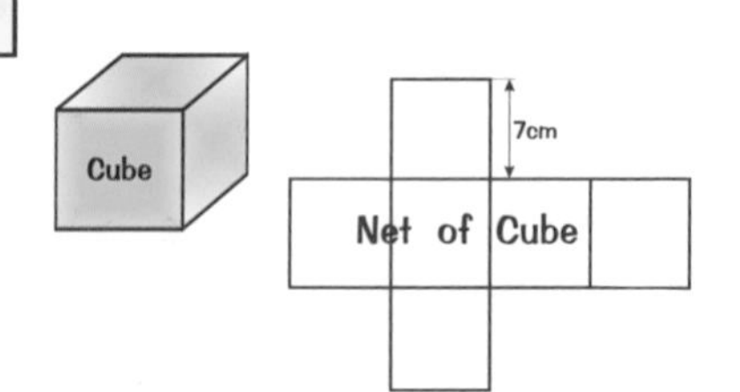

3) Cuboid

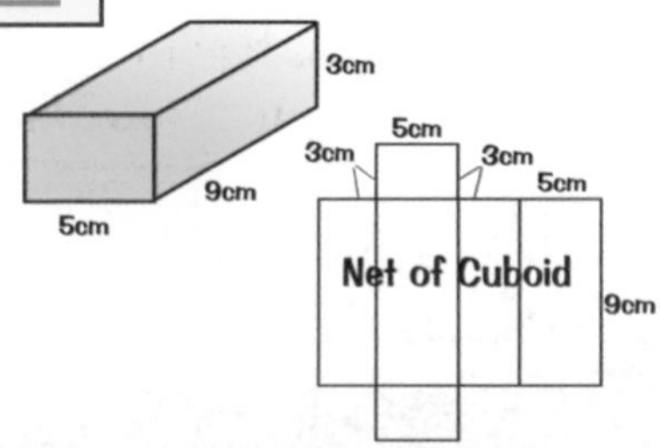

4) Pyramid

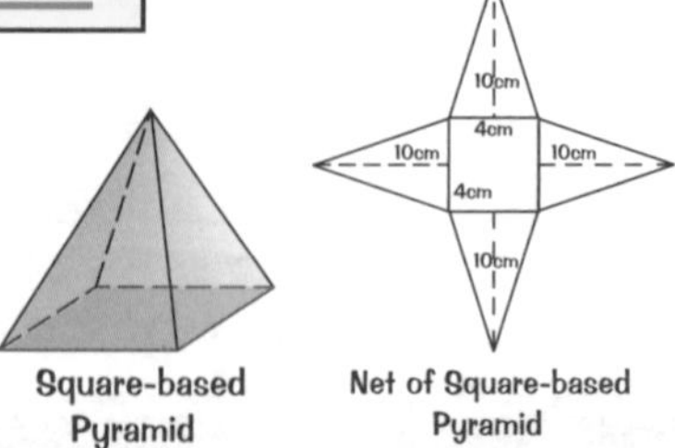

The Acid Test:

LEARN the formulas for Areas of Rectangles and Triangles, the 4 details on Surface Area and Nets, and the FOUR NETS on this page.

Now cover the page and write down everything you've learnt.
Work out the area of all four nets shown above.

Volume or Capacity

VOLUMES — YOU MUST LEARN THESE TOO!

1) CUBOID (RECTANGULAR BLOCK)

(This is also known as a 'rectangular prism' — see below to understand why)

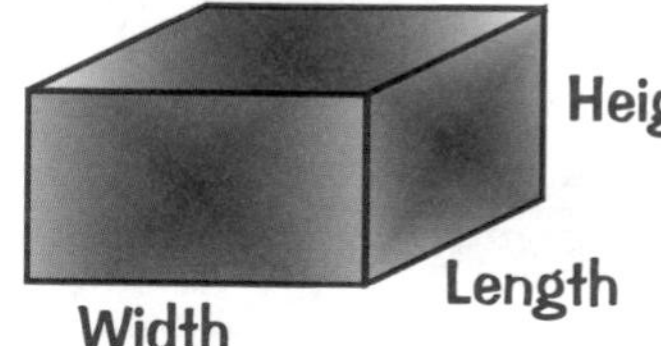

Volume of Cuboid = length × width × height

(The other word for volume is CAPACITY)

2) PRISM

A PRISM is a solid (3-D) object which has a CONSTANT AREA OF CROSS-SECTION — i.e. it's the same shape all the way through.

Now, for some reason, not a lot of people know what a prism is, but they come up all the time in Exams, so make sure YOU know.

Length

Constant Area of Cross-section

Volume of prism = Cross-sectional Area × length

V = A × l

Circular Prism (or Cylinder)

Constant Area of Cross-section

Length

Triangular Prism

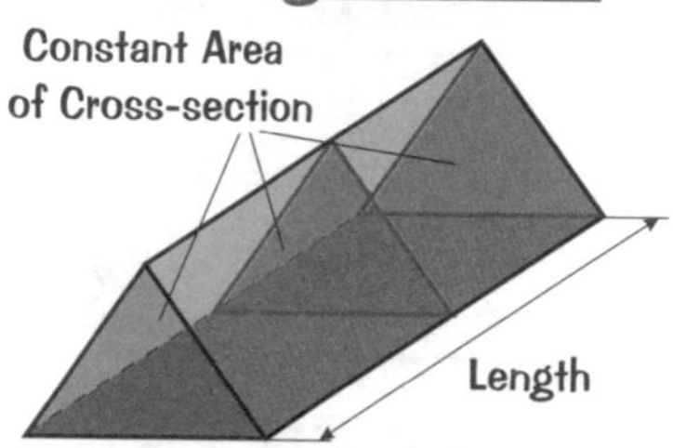

As you can see, the formula for the volume of a prism is very simple. The difficult part, usually, is finding the area of the cross-section.

The Acid Test:

LEARN this page. Then turn over and try to write it all down. Keep trying until you can do it.

1) Practise these two questions until you can do them both without any hesitation. Name the shapes and find their volumes:

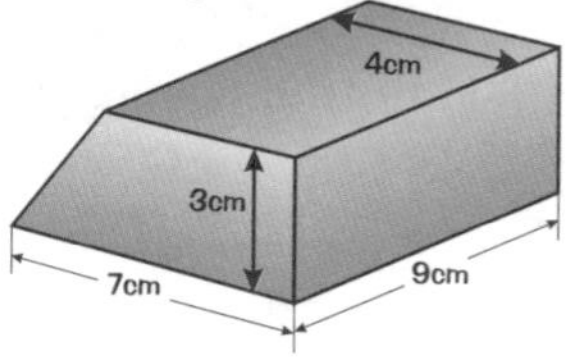

b)

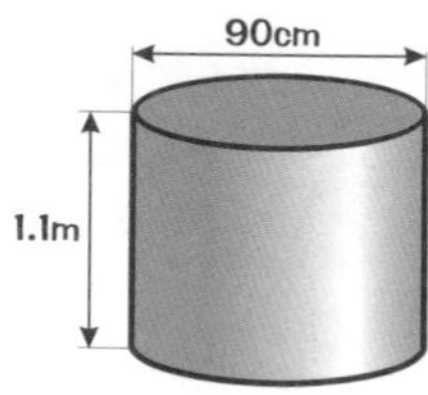

Geometry

7 Simple Rules — that's all:

If you know them all — thoroughly, you at least have a fighting chance of working out problems with lines and angles. If you don't — you've no chance.

1) Angles in a triangle

Add up to 180°.

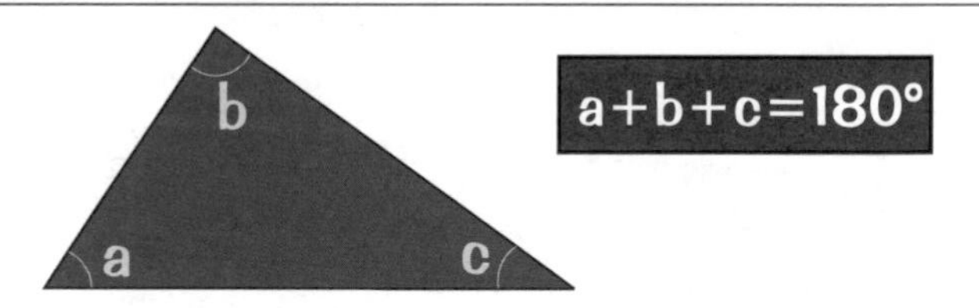

2) Angles on a straight line

Add up to 180°.

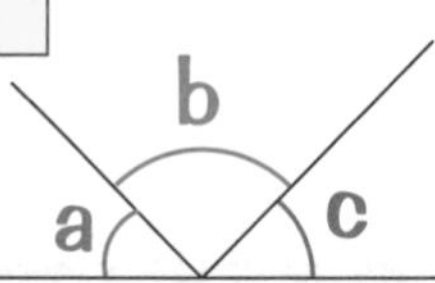

a+b+c=180°

3) Angles in a 4-sided shape

(a "Quadrilateral")

Add up to 360°.

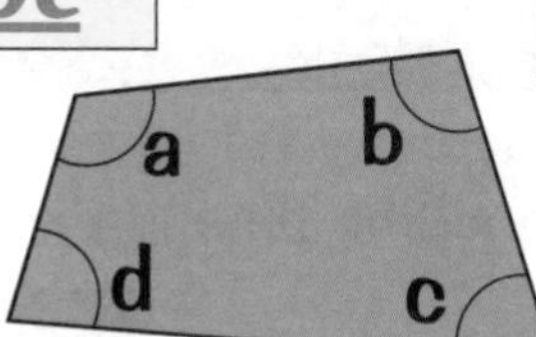

a+b+c+d=360°

4) Angles round a point

Add up to 360°.

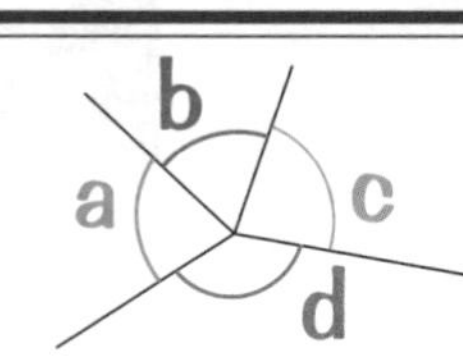

a+b+c+d=360°

5) Exterior Angle of Triangle

Exterior Angle of triangle
= sum of Opposite Interior angles

i.e. a+b=d

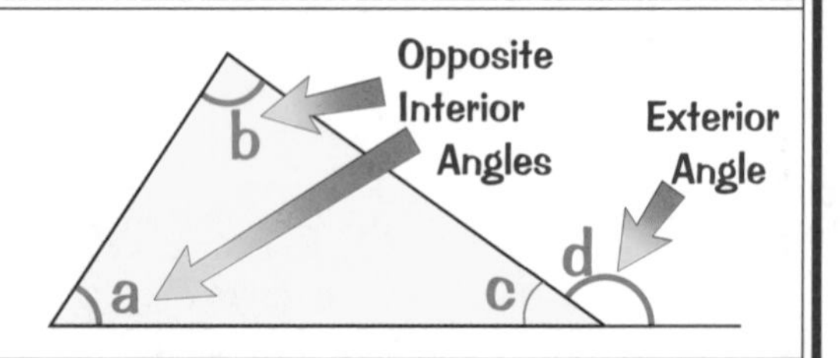

6) Isosceles triangles

2 sides the same
2 angles the same

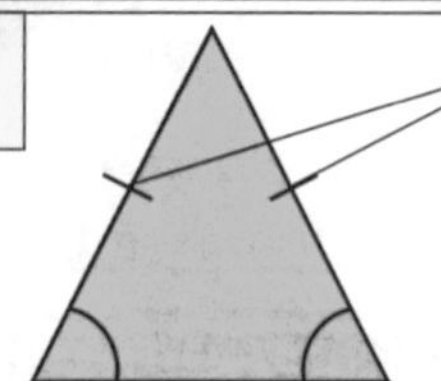

These dashes indicate two sides the same length

In an isosceles triangle, you only need to know one angle to be able to find the other two, which is very useful if you remember it.

a)

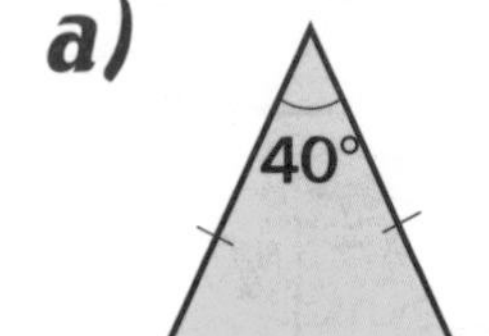

180° – 40° = 140°
The two bottom angles are both the same and they must add up to 140°, so each one must be half of 140° (= 70°). So X = 70°.

b)

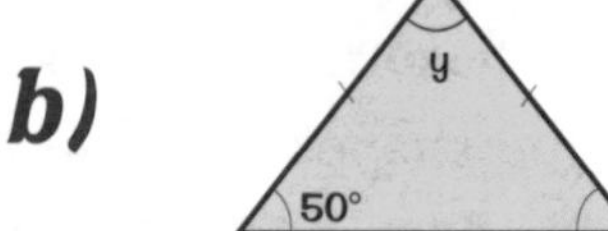

The two bottom angles must be the same, so 50° + 50° = 100°.
All the angles add up to 180° so Y = 180° – 100° = 80°.

Geometry

7) Parallel Lines

Whenever one line crosses two parallel lines then the two bunches of angles are the same, and $a + b = 180°$

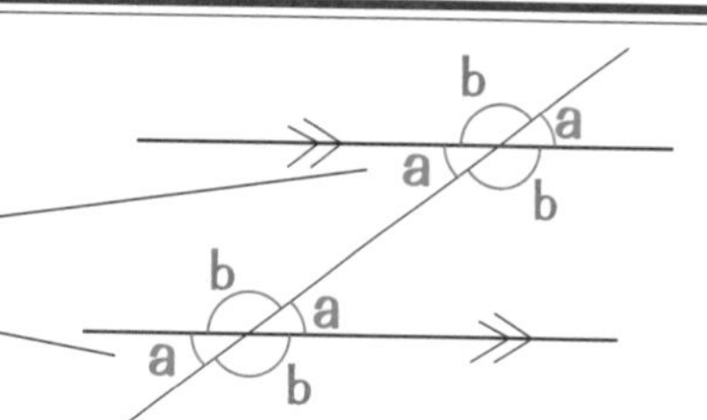

You need to spot the characteristic Z, C, U and F shapes:

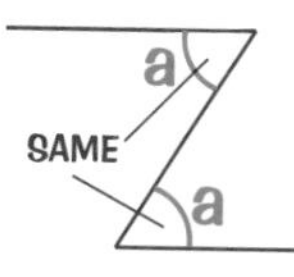

In a Z-shape they're called "ALTERNATE ANGLES"

a

ADD UP TO 180

b

If they add up to 180 they're called "SUPPLEMENTARY ANGLES"

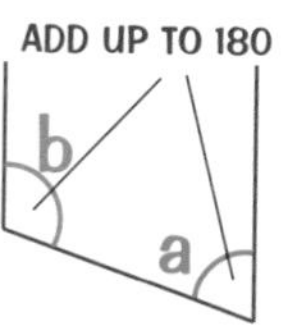

a

a

SAME

In an F-shape they're called "CORRESPONDING ANGLES"

Alas you're expected to learn these three silly names too!

If necessary, EXTEND THE LINES to make the diagram easier to get to grips with:

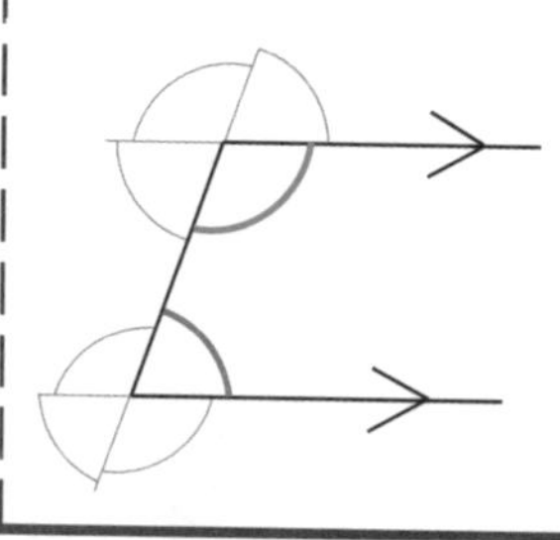

The Basic Approach to Geometry Problems

1) Don't concentrate too much on the angle you have been asked to find. The best method is to find ALL the angles in whatever order they become obvious.

2) Don't sit there waiting for inspiration to hit you. It's all too easy to find yourself staring at a geometry problem and getting nowhere. The method is this:

GO THROUGH ALL THE ABOVE RULES OF GEOMETRY, ONE BY ONE, and apply each of them in turn in as many ways as possible — one of them is bound to work.

An Example

"Find all the other angles in this diagram."

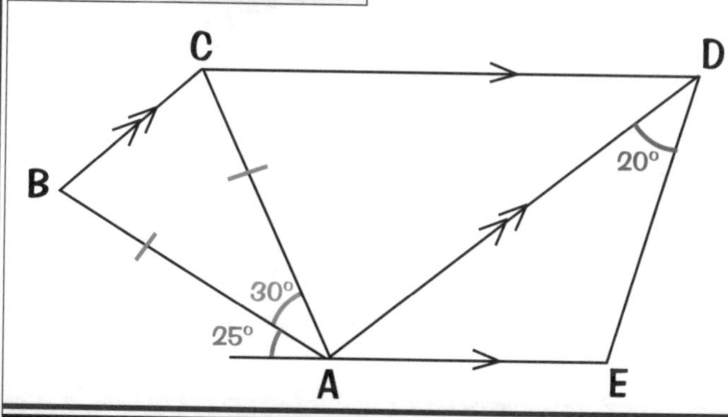

ANSWER:

1) ABC is isosceles, so $\angle ABC = \angle ACB = 75°$
2) BC and AD are parallel, BCAD is a Z-shape, so if $\angle ACB = 75°$ then $\angle CAD = 75°$ too.
3) Angles on a straight line means $\angle EAD = 50°$
4) AE and CD are parallel so $\angle ADC = 50°$ also.
5) Triangle ACD adds up to 180° so $\angle ACD = 55°$
6) Triangle ADE adds up to 180° so $\angle AED = 110°$

The Acid Test:

LEARN EVERYTHING on these 2 pages, especially all the stuff on parallel lines, then turn over and write it all down.

1) If one angle of an isosceles triangle is 68°, what values could the other angles have?
2) Find angle x in this diagram and then fill in all the other angles.

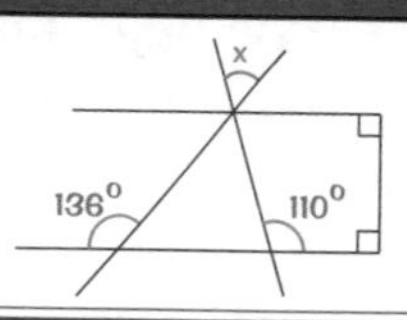

Circles

Now it's time for some lovely Geometry. Learn this page about Circles.

1) Don't muddle up these Two Circle Formulas

AREA of CIRCLE = $\pi \times (\text{radius})^2$

$$A = \pi \times r^2$$

$\pi = 3.141592....$
$= 3.14$ (approx)

e.g. if the radius is 4cm, then
$A = 3.14 \times (4 \times 4)$
$= 50.24\text{cm}^2$

Radius

Diameter

CIRCUMFERENCE = $\pi \times$ diameter

$$C = \pi \times D$$

Circumference = distance round the outside of the circle

2) Arc, Chord and Tangent

A TANGENT is a straight line that just touches the outside of the circle.

A CHORD is a line drawn across the inside of a circle.

AN ARC is just part of the circumference of the circle.

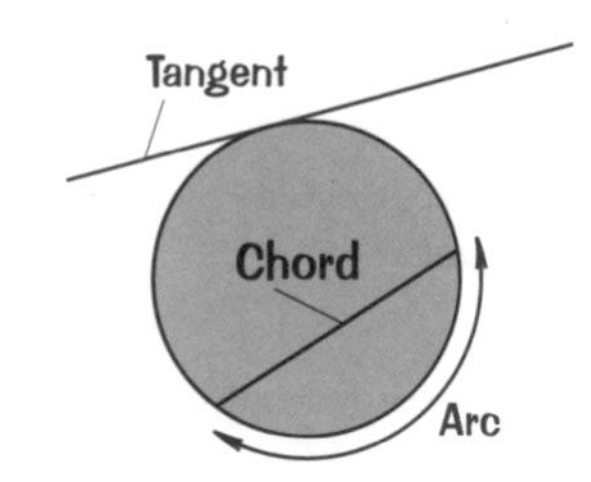

3) Sectors and Segments are both Areas

SECTOR OF CIRCLE

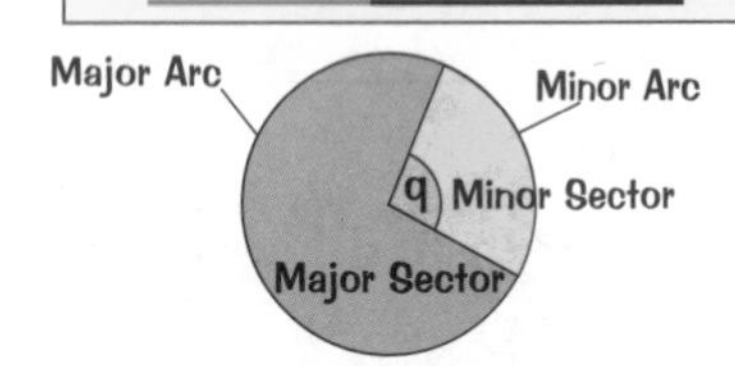

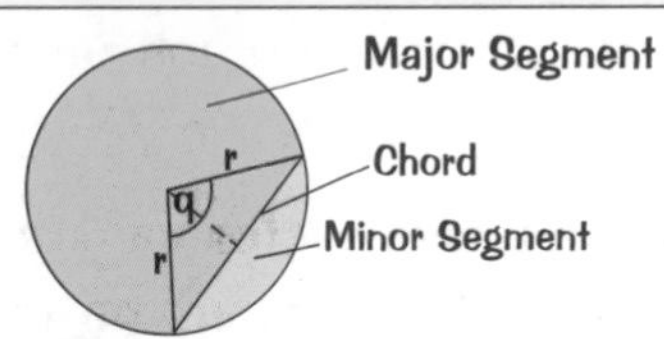

4) Tangent and Radius Meet at 90°

A TANGENT is a line that just touches the edge of a circle (or other curve). If a tangent and radius meet at the same point, then the angle they make is EXACTLY 90°.

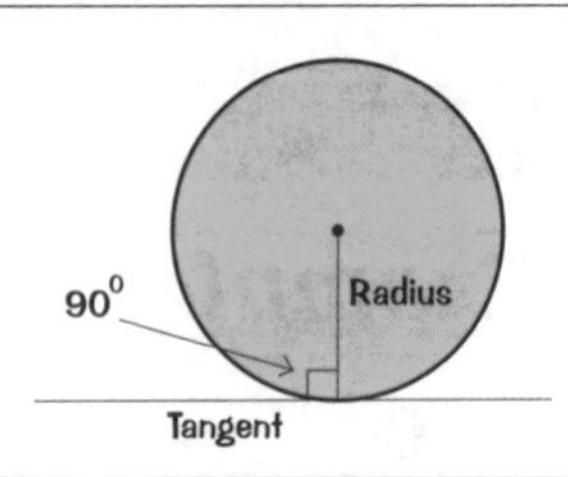

5) Equality of Tangents from a Point

The two tangents drawn from an outside point are always equal in length, so creating an "isosceles" situation, with two congruent right-angled triangles.

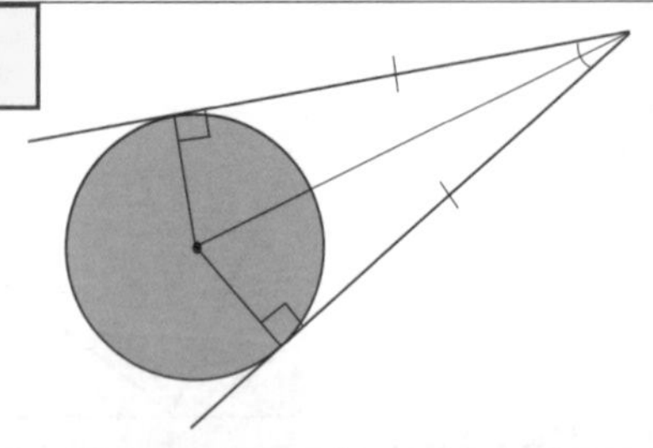

The Acid Test:

There are 5 SECTIONS on this page.
They're all mighty important — LEARN THEM.

Now cover the page and write down everything you've learnt. Frightening isn't it.

1) A plate has a diameter of 14cm. Find the area and the circumference of the plate. Remember to show your working out.
2) A flower bed has a radius of 6m. Find the area and circumference of it.

X, Y and Z Coordinates

The Four Quadrants

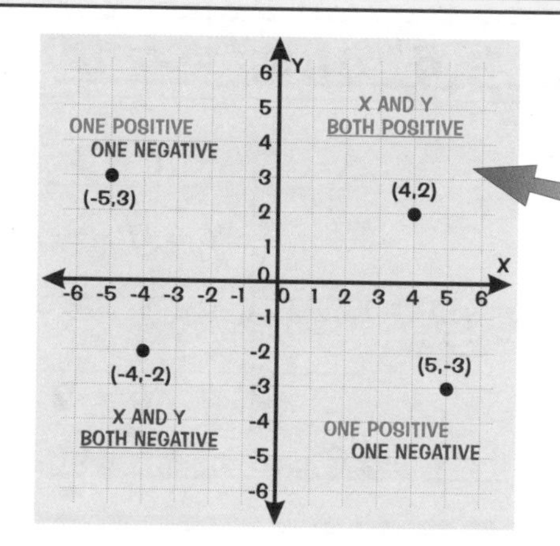

A graph has four different quadrants (regions) where the x- and y- coordinates are either positive or negative.

This is the easiest region by far because here all the coordinates are positive.

You have to be careful in the other regions though, because the x- and y- coordinates could be negative, and that always makes life much more difficult.

Coordinates are always written in brackets like this: (x, y) — remember x is across, and y is up.

Finding the Midpoint of a Line Segment

This regularly comes up in exams and is dead easy...

1) Find the average of the two x-coordinates, then do the same for the y-coordinates.
2) These will be the coordinates of the midpoint.

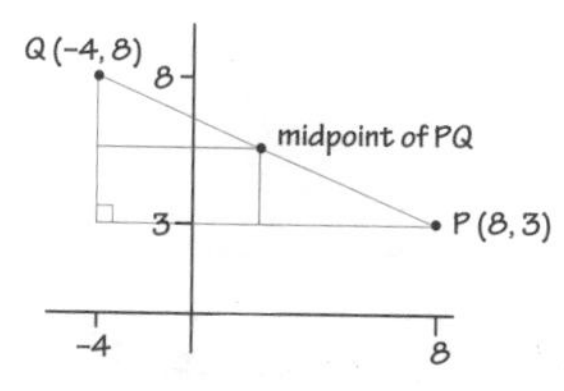

Example:

"Point P has coordinates (8, 3) and point Q has coordinates (-4, 8). Find the midpoint of the line PQ."

Solution:

Average of x-coordinates = (8 + -4)/2 = 2
Average of y-coordinates = (8 + 3)/2 = 5.5
So, coordinates of midpoint = (2, 5.5)

Z Coordinates are for 3-D space

1) All z-coordinates do is extend the normal x-y coordinates into a third direction, z, so that all positions then have 3 coordinates: (x,y,z)

2) This means you can give the coordinates of the corners of a box or any other 3-D SHAPE.

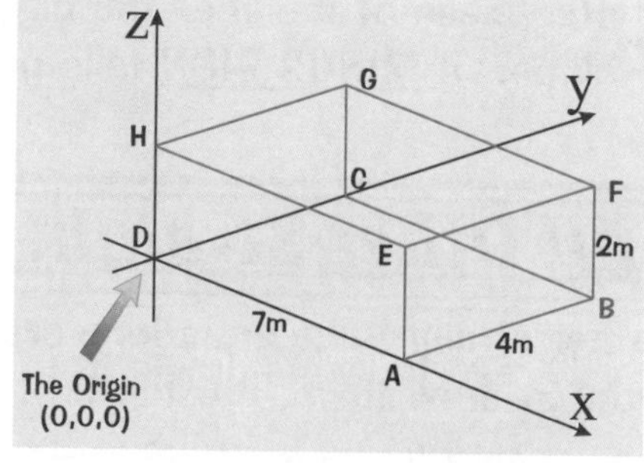

For example in this drawing, the coordinates of B and F are B(7,4,0) F(7,4,2)

The Acid Test:

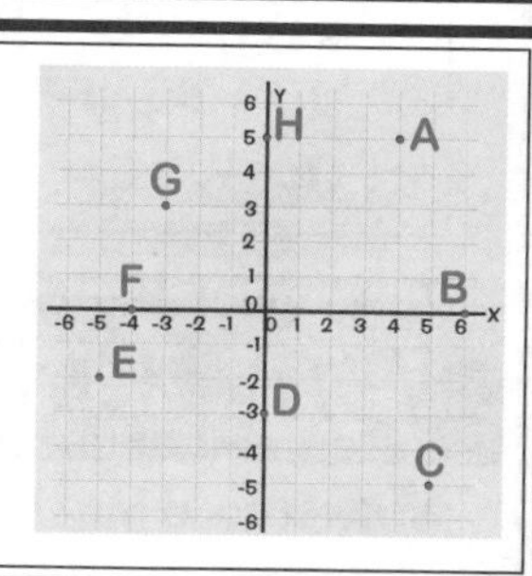

1) Write down the coordinates of the letters A to H on this graph:
2) Find the coordinates of the points exactly midway between:
   a) A and B b) F and H c) E and C
3) Find the coordinates of all the points on the 3D diagram above.

Straight Line Graphs

Any straight line graph can be described by a simple equation.
You should be able to recognise a lot of graphs just from their equations.

1) Horizontal and Vertical lines: "x = a" and "y = b"

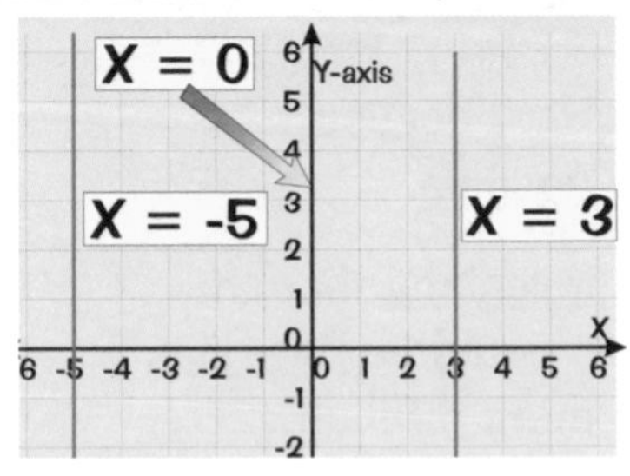

x = a is a vertical line through "a" on the x-axis

y = a is a horizontal line through "a" on the y-axis

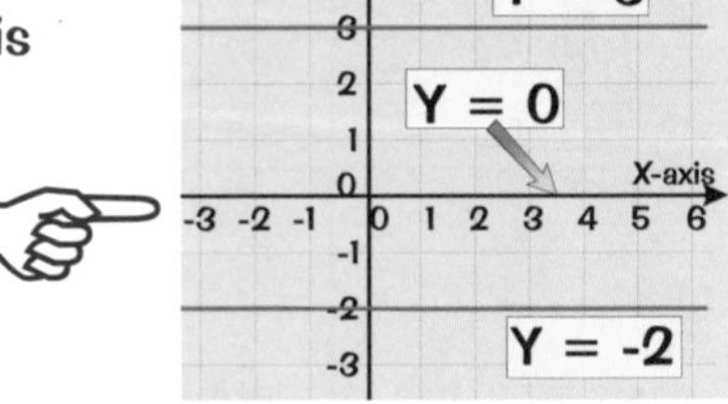

Don't forget: the y-axis is also the line x=0

Don't forget: the x-axis is also the line y=0

2) The Main Diagonals: "y = x" and "y = –x"

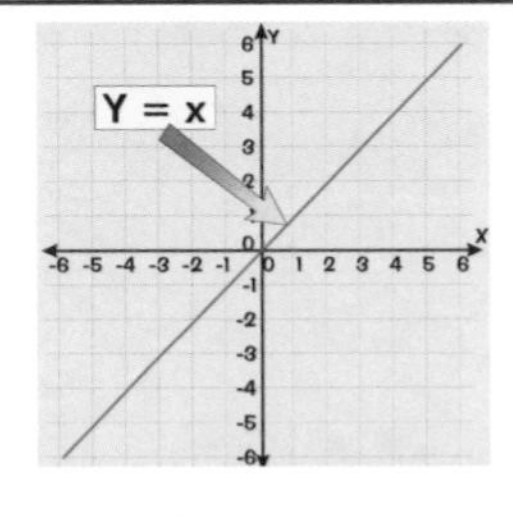

"y = x" is the main diagonal that goes UPHILL from left to right.

"y = -x" is the main diagonal that goes DOWNHILL from left to right.

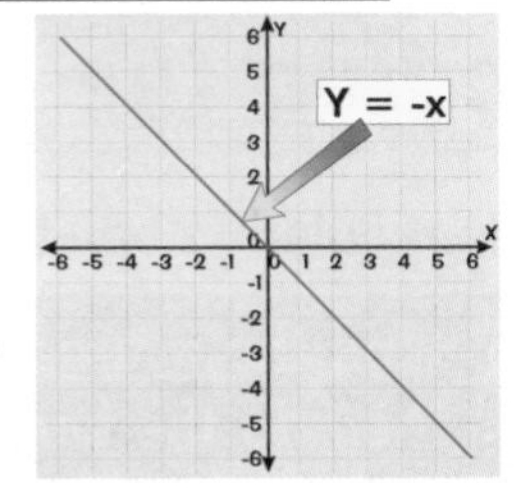

3) Other Sloping Lines Through the origin: "y = ax" and "y = –ax"

y = ax and y = -ax are the equations for A SLOPING LINE THROUGH THE ORIGIN.

The value of "a" (known as the gradient) tells you the steepness of the line. The bigger "a" is, the steeper the slope. A MINUS SIGN tells you it slopes DOWNHILL.

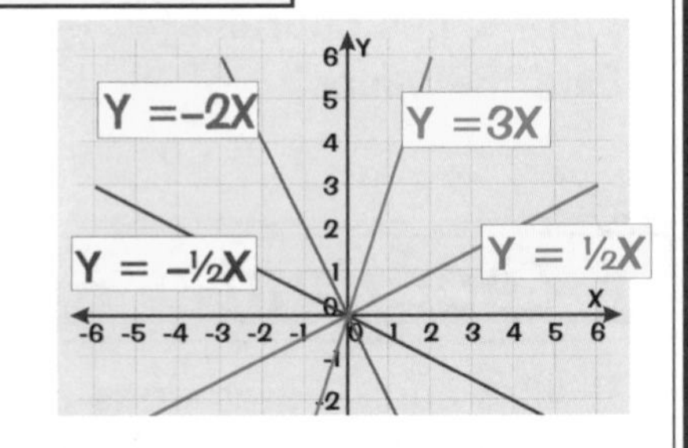

All Other Straight Lines

Other straight-line equations are a little more complicated. The next page shows you how to draw them, but the first step is identifying them in the first place. Remember: All straight line equations just contain "something x, something y, and a number".

Straight lines:

$x - y = 0$	$y = 2 + 3x$
$2y - 4x = 7$	$4x - 3 = 5y$
$3y + 3x = 12$	$6y - x - 7 = 0$

NOT straight lines:

$y = x^3 + 3$	$2y - 1/x = 7$
$1/y + 1/x = 2$	$x(3 - 2y) = 3$
$x^2 = 4 - y$	$xy + 3 = 0$

The Acid Test:

LEARN all the specific graphs on this page and also how to identify straight line equations.

Now turn over the page and write down everything you've learned.

Straight Line Graphs

In the exam you'll be expected to draw the graphs of straight line equations. This page shows you two easy methods.

1) The "Table of 3 values" method

You can easily draw the graph of any equation using this easy method:

1) Choose 3 values of x and draw up a wee table,
2) Work out the y-values,
3) Plot the coordinates, and draw the line.

If it's a straight line equation, the 3 points will be in a dead straight line with each other, which is the usual check you do when you've drawn it — if they aren't, then it could be a curve and you'll need to do more values in your table to find out what on earth's going on.

Example: "Draw the graph of $y = 2x - 3$"

1) Draw up a table with some suitable values of x. Choosing $x = 0, 2, 4$ is usually cool enough. i.e.

X	0	2	4
Y			

2) Find the y-values by putting each x-value into the equation:

X	0	2	4
Y	-3	1	5

(e.g. When $x = 4$, $y = 2x - 3 = 2 \times 4 - 3 = 5$)

3) Plot the points and draw the line.

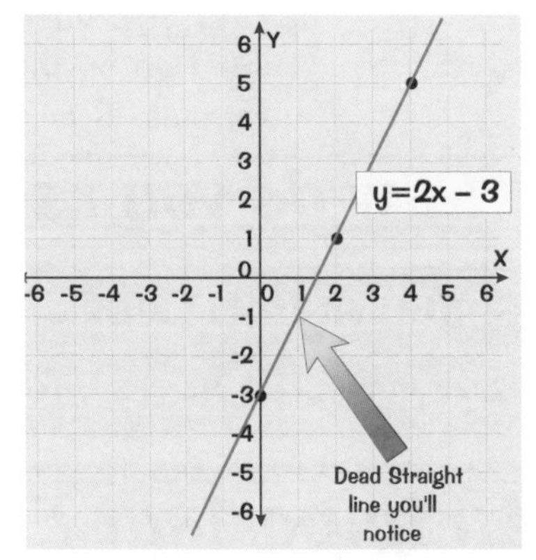

2) The "x = 0", "y = 0" method

1) Set x=0 in the equation, and find y — this is where it crosses the y-axis.
2) Set y=0 in the equation and find x — this is where it crosses the x-axis.
3) Plot these two points and join them up with a straight line — and just hope it should be a straight line, since with only 2 points you can't really tell, can you!

Example: "Draw the graph of $5x + 3y = 15$"

1) Putting $x = 0$ gives "$3y = 15$" $\Rightarrow$ $y = 5$
2) Putting $y = 0$ gives "$5x = 15$" $\Rightarrow$ $x = 3$
3) So plot (0, 5) and (3, 0) on the graph and join them up with a straight line:

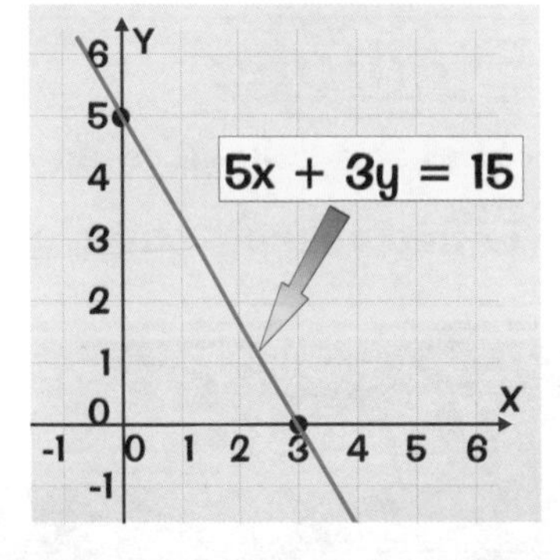

Only doing 2 points is risky unless you're sure the equation is definitely a straight line — but then that's the big thrill of living life on the edge, isn't it.

The Acid Test:

LEARN the details of these TWO EASY METHODS then turn over and write down all you know.

1) Draw these graphs using both methods a) $y = 4 + x$ b) $4y + 3x = 12$ c) $y = 6 - 2x$

Formula Triangles

You may have already come across these in physics, because they are extremely potent tools for dealing swiftly and reliably with a lot of common formulas. They are very easy, so make sure you know how to use them.

Where do you put the letters?

If 3 things are related by a formula like this: $A = B \times C$ or like this: $B = A/C$
then you can put them into a FORMULA TRIANGLE thus:

1) $A = B \times C$ If there are TWO LETTERS MULTIPLIED TOGETHER they must go ON THE BOTTOM of the Formula Triangle, and so the other must go on the top. For example the formula $A = B \times C$ becomes:

2) $B = A/C$ If there is ONE THING DIVIDED BY ANOTHER then the one ON TOP OF THE DIVISION goes ON TOP IN THE FORMULA TRIANGLE, and so the other two letters must go on the bottom (it doesn't matter which way round). For example, the formula $B = A/C$ will produce the same formula triangle as the one above.

How do you use it?

1) COVER UP the thing you want to find and just WRITE DOWN what is left showing.

2) Now PUT IN THE VALUES for the other two things and WORK IT OUT.

An Important Example:

DENSITY = MASS ÷ VOLUME

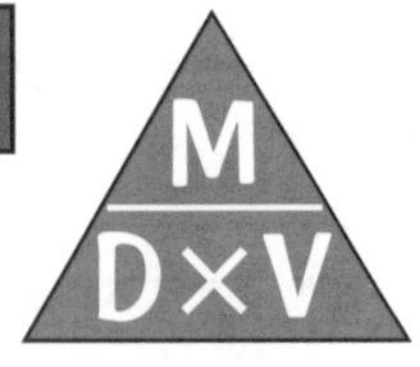

The standard formula for density is: Density = Mass / Volume
so we can put it in a FORMULA TRIANGLE like this:

You might think this is physics, but density is specifically mentioned in the maths syllabus.

One way or another you MUST remember this formula for density, because they promise nothing and without it you'll be stuck. The best method by far is to remember the order of the letters in the formula triangle as D^MV or DiMoV (The Russian Agent!).

Example:

"Find the volume of an object with a mass of 40g and a density of $6.4 g/cm^3$."

To find volume, cover up V. This leaves M/D, so $V = M \div D = 40 \div 6.4 = 6.25 \text{ cm}^3$.

The Acid Test:

LEARN the 2 Rules for creating Formula Triangles, the 2 Rules for Using Them, and the Density Formula.
Then turn over and write it all down from memory.

1) A metal object has a volume of $45 cm^3$ and a mass of 743g. What is its density?
2) Another piece of the same metal has a volume of $36.5 cm^3$. What is its mass?
3) What are the formula triangles for a) $F = m \times a$ b) $P = F/A$ c) $V = I \times R$

Speed, Distance and Time

This is very common, and they never give you the formula! Either you learn it beforehand or you wave goodbye to several easy marks.

1) The Formula Triangle

Of course you have to remember the order of the letters in the triangle (SDT), and this time we have the word SoDiT to help you. So if it's a question on speed, distance and time just say: **SOD IT**.

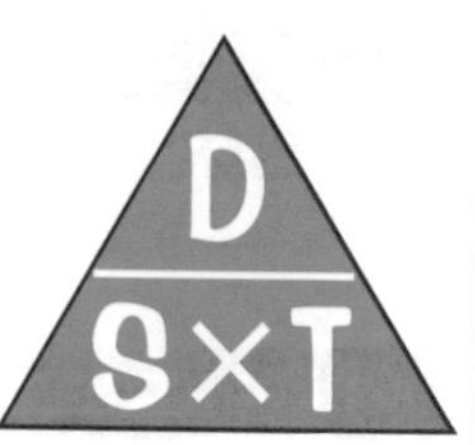

EXAMPLE: "A car travels 90 miles at 36 miles per hour. How long does it take?"

ANS: We want to find the time, so cover up T in the triangle which leaves D/S, so T = D/S = Distance ÷ speed = 90 ÷ 36 = 2.5 hours

IF YOU LEARN THE FORMULA TRIANGLE, YOU WILL FIND QUESTIONS ON SPEED, DISTANCE AND TIME VERY EASY.

2) Units — Getting them right

By units we mean things like cm, m, m/s, km^2, etc. and quite honestly they should always be in your mind when you write an answer down. When you're using a FORMULA, there is one special thing you need to know. It's simple enough but you must know it:

The UNITS you get out of a Formula DEPEND ENTIRELY upon the UNITS you put into it.

For example if you put a distance in cm and a time in seconds into the formula triangle to work out SPEED, the answer must come out in cm per second (cm/s).

If the time is in hours and the speed in miles per hour (mph) then the distance you'd calculate would come out in miles. It's pretty simple when you think about it.

But Don't Mix Units

E.g. Don't mix Miles Per HOUR in a formula with a time in MINUTES (convert it to hours).
Don't mix DENSITY IN g/cm^3 in a formula with a MASS IN kg (convert it to g).

Example:

"A boy walks 800m in 10minutes. Find his speed in km/h."

If you use 800m and 10 minutes your answer will be a speed in metres per minute (m/min). Instead you must convert: 800m = 0.8 km, 10 mins = 0.1667 hours (mins÷60). Then you can divide 0.8 km by 0.16667 hours to get 4.8 km/h.

The Acid Test:

LEARN THIS PAGE, then turn over and briefly summarise every topic, with examples. Keep trying until you can.

1) Find the time taken, in hours, mins and secs, for a purple-nosed buffalo walking at 3.2 km/h to cover 5.2km.
2) Also find how far it would go in 35 mins and 25 seconds. Give it in both km and m.

Accuracy and Estimating

Appropriate Accuracy

To decide what is appropriate accuracy, you need only remember these three rules:

1) For fairly casual measurements, 2 SIGNIFICANT FIGURES is most appropriate.

EXAMPLES: Cooking – 250g (2 sig fig) of sugar, not 253g (3 S F), or 300g (1 S F)
Distance of a journey – 450 miles or 25 miles or 3500 miles (All 2 S F)
Area of a garden or floor — $330m^2$ or $15m^2$

2) For more important or technical things, 3 SIGNIFICANT FIGURES is essential.

EXAMPLES: A technical figure like 34.2 miles per gallon, rather than 34 mpg.
A length that is cut to fit, e.g. measure a shelf 25.6cm long not just 26cm.
Any accurate measurement with a ruler: 67.5cm not 70cm or 67.54cm

3) Only for really scientific work would you have more than 3 SIG FIG.

Estimating

This is VERY EASY, so long as you don't over-complicate it.

1) ROUND EVERYTHING OFF to nice easy CONVENIENT NUMBERS
2) Then WORK OUT THE ANSWER using these nice easy numbers — that's it!

In the Exam you'll need to show all the steps, to prove you didn't just use a calculator.

EXAMPLE: Estimate the value of $\frac{127.8+41.9}{56.5\times3.2}$ showing all your working.

Ans: $\frac{127.8+41.9}{56.5\times3.2} \approx \frac{130+40}{60\times3} \approx \frac{170}{180} \approx 1$ (" ≈ " means "roughly equal to")

Estimating Areas and Volumes

1) Draw or imagine a RECTANGLE OR CUBOID of similar size to the object in question
2) Round off all lengths to the NEAREST WHOLE, and work it out — easy.

EXAMPLES: "Estimate the area of this shape and the volume of the bottle:"

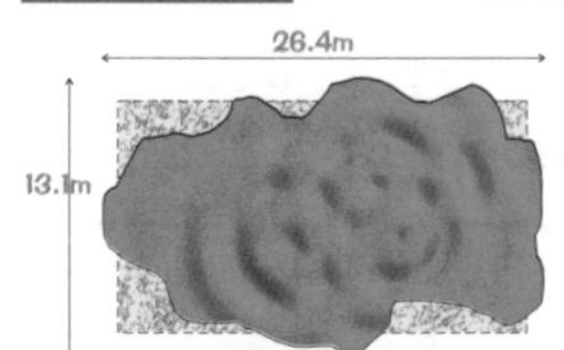

Area ≈ rectangle
26m × 13m = $338m^2$
(or without a calculator:
30 x 10 = $300m^2$)

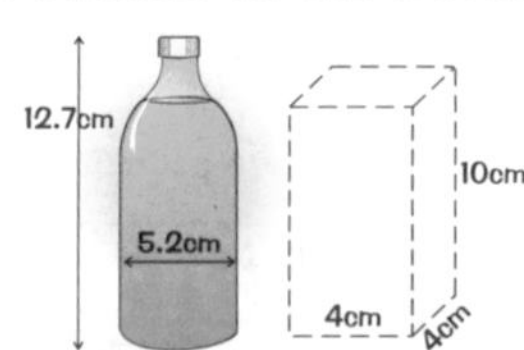

Volume ≈ cuboid
=4 × 4 × 10
=$160cm^3$

Estimating Square Roots

1) Find the TWO SQUARE NUMBERS EITHER SIDE of the number in question.
2) Find the SQUARE ROOTS and pick a SENSIBLE NUMBER IN BETWEEN.

EXAMPLE: "Estimate $\sqrt{85}$ without using a calculator."

① The square numbers either side of 85 are 81 and 100.

② The square roots are 9 and 10, so $\sqrt{85}$ must be between 9 and 10. But 85 is much nearer 81 than 100, so $\sqrt{85}$ must be much nearer 9 than 10. So pick 9.1, 9.2 or 9.3. (The answer's actually 9.2195... if you're interested.)

The Acid Test:

LEARN the 3 Rules for Appropriate Accuracy and 6 Rules for Estimating. Then turn over and write them all down.

1) Decide which category of accuracy these should belong in and round them off accordingly:
a) A jar of jam weighs 34.56g b) A car's max speed is 134.25mph c) A cake needs 852.3g of flour
2) Estimate the area of Great Britain in Square miles, and the volume of a tin of beans in cm^3.
3) Without your calculator, estimate: a) $\sqrt{12}$, b) $\sqrt{104}$, c) $\sqrt{52}$ d) $\sqrt{30}$.

Unit Three Revision Summary

These questions may seem difficult, but they are the very best revision you can do. They follow the sequence of pages in Unit Three, so you can easily look up anything you don't know.

Keep learning these basic facts until you know them

1) What are square numbers and cube numbers?
2) List the first ten triangle numbers.
3) What is the method for deciding if a number is prime?
4) What are factors?
5) What are the LCM and HCF?
6) How do you convert from fractions to decimals to percentages and back again?
7) Write down the 2 formulas for finding the nth term of a number pattern.
8) Give an example of when the memory buttons on your calculator should be used.
9) Explain what BODMAS is. Does your calculator know about it?
10) Give a good example of when the brackets buttons should be used.
11) How do you enter a time in hours, minutes and seconds on your calculator?
12) How do you convert hours, minutes and seconds to decimal time and back again?
13) What are the three steps for using conversion factors? Give three examples.
14) Give 8 metric, 5 imperial and 8 metric-to-imperial conversions.
15) What's the method for writing a number to three significant figures?
16) What is a term? Give an example.
17) Describe the steps for simplifying an expression involving several terms.
18) What does "D.O.T.S." stand for? Give two examples of it.
19) What is the method for multiplying out double brackets and squared brackets?
20) What are the rules for multiplying, dividing, adding and subtracting algebraic fractions?
21) Write down the ten rules for powers and roots.
22) What is the format of any number expressed in standard form?
23) Which is the standard form button? What would you press to enter 6×10^8?
24) Describe the method for multiplying or dividing numbers written in standard form.
25) What are the formulas for the areas of triangles and rectangles?
26) How are shape nets related to surface area?
27) What are the seven geometry rules you need to know?
28) What do the following circle terms mean — show them on a diagram.
    Radius, diameter, circumference, arc, chord, segment, sector, tangent.
29) How do you find the midpoint of a line segment?
30) How can you recognise the equation of a straight line.
31) How would you draw a straight line from its equation?
32) Write down the formula triangles for speed and density. Explain how you use them.
33) Give three rules for deciding on appropriate accuracy.
34) How do you estimate a square root?

Manipulating Surds and Use of π

RATIONAL NUMBERS The vast majority of numbers are rational. They are always either:

1) A whole number (either positive (+ve), or negative (–ve)), e.g. 4, -5, -12
2) A fraction p/q, where p and q are whole numbers (+ve or –ve), e.g. ¼, -½, ¾
3) A terminating or recurring decimal, e.g. 0.125, 0.3333333333..., 0.143143143143...

IRRATIONAL NUMBERS are messy!

1) They are always never-ending non-repeating decimals. π is irrational.
2) A good source of irrational numbers is square roots and cube roots.

Manipulating Surds

It sounds like something to do with controlling difficult children, but it isn't. Surds are expressions with irrational square roots in them. You MUST USE THEM if they ask you for an EXACT answer. There are a few simple rules to learn:

1) $\sqrt{a} \times \sqrt{b} = \sqrt{ab}$ e.g. $\sqrt{2} \times \sqrt{3} = \sqrt{2 \times 3} = \sqrt{6}$ — also $(\sqrt{b})^2 = b$, fairly obviously
2) $\sqrt{a}/\sqrt{b} = \sqrt{a/b}$ e.g. $\sqrt{8}/\sqrt{2} = \sqrt{8/2} = \sqrt{4} = 2$
3) $\sqrt{a} + \sqrt{b}$ — NOTHING DOING... (in other words it is definitely NOT $\sqrt{a+b}$)
4) $(a + \sqrt{b})^2 = (a + \sqrt{b})(a + \sqrt{b}) = a^2 + 2a\sqrt{b} + b$ (NOT just $a^2 + (\sqrt{b})^2$)
5) $(a + \sqrt{b})(a - \sqrt{b}) = a^2 + a\sqrt{b} - a\sqrt{b} - (\sqrt{b})^2 = a^2 - b$
6) Express $3/\sqrt{5}$ in the form $a\sqrt{5}/b$ where a and b are whole numbers.
   To do this you must "RATIONALISE the denominator", which just means multiplying top and bottom by $\sqrt{5}$: $3\sqrt{5}/\sqrt{5}\sqrt{5} = 3\sqrt{5}/5$ so a = 3 and b = 5
7) If you want an exact answer, LEAVE THE SURDS IN. As soon as you go using that calculator, you'll get a big fat rounding error — and you'll get the answer WRONG. Don't say I didn't warn you...

Example: A square has an area of 15 cm². Find the length of one of its sides.
Answer: The length of a side is $\sqrt{15}$ cm.
If you have a calculator, then you can work out $\sqrt{15}$ = 3.8729833...cm.
If you're working without a calculator, or are asked to give an EXACT answer, then just write: $\sqrt{15}$ cm. That's all you have to do.

Exact calculations using π — Leave π in the answer

π is an irrational number that often comes up in calculations, e.g. in finding the area of a circle. Most of the time you can use the nifty little π button on your calculator. But if you're asked to give an exact answer or, worse still, do the calculation without a calculator, just leave the π symbol in the calculation.

Example: Find the area of a circle with radius 4 cm, without using a calculator.
Answer: The area $= \pi r^2 = \pi \times 4^2 = 16\pi$ cm².

The Acid Test:

LEARN the 7 rules for manipulating surds, then turn over and write them all down.

Simplify 1) $(1+\sqrt{2})^2 - (1-\sqrt{2})^2$ 2) $(1+\sqrt{2})^2 - (2\sqrt{2} - \sqrt{2})^2$

Upper Bounds and Reciprocals

This page covers some lovely bits and bobs which, well frankly, didn't fit anywhere else. That doesn't mean they're not important though. In fact, upper / lower bound calculations are a favourite with examiners, so make sure you learn all these details...

1) Finding the Upper and Lower bounds of a Single Measurement

The simple rule is this: **The real value can be as much as HALF THE ROUNDED UNIT above and below the rounded-off value**

E.g. If a length is given as 2.4 m to the nearest 0.1 m, the rounded unit is 0.1 m so the real value could be anything up to 2.4 m ± 0.05 m giving answers of 2.45 m and 2.35 m for the upper and lower bounds.

2) The Maximum and Minimum Possible Values of a Calculation

When a calculation is done using rounded-off values there will be a DISCREPANCY between the CALCULATED VALUE and the ACTUAL VALUE:

EXAMPLE: A floor is measured as being 5.3 m × 4.2 m to the nearest 10 cm. This gives an area of 22.26 m^2, but this is not the actual floor area because the real values could be anything from 5.25 m to 5.35 m and 4.15 m to 4.25 m,

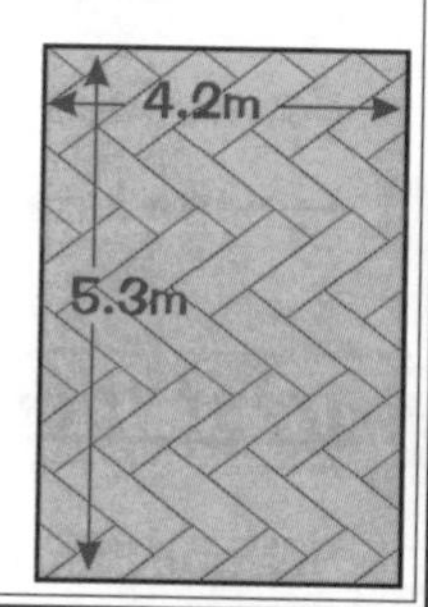

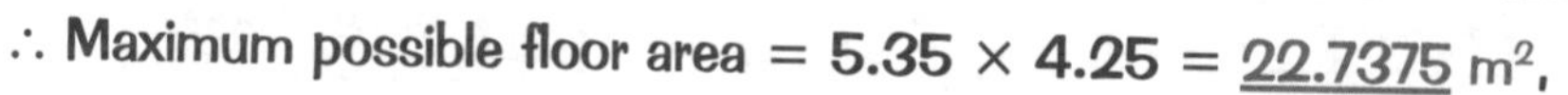

∴ Maximum possible floor area = 5.35 × 4.25 = 22.7375 m^2,

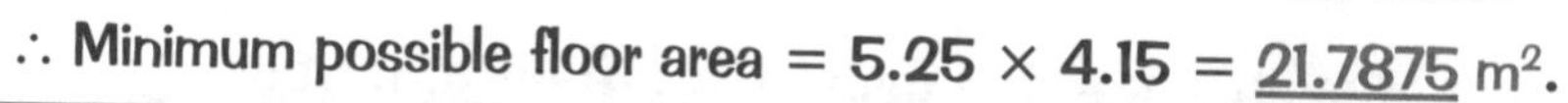

∴ Minimum possible floor area = 5.25 × 4.15 = 21.7875 m^2.

Reciprocals — Learn these 4 Facts

1) The reciprocal of a number is "one over" the number. **The reciprocal of 5 = $\frac{1}{5}$.**
2) You can find the reciprocal of a fraction by turning it upside down. **The reciprocal of $\frac{3}{8} = \frac{8}{3}$.**
3) A number multiplied by its reciprocal gives 1. **$\frac{6}{7} \times \frac{7}{6} = 1$**
4) 0 has no reciprocal because you can't divide anything by 0.

The [1/x] (or [x^{-1}]) Button Makes Reciprocals much Easier

This has two very useful functions:

1) Making divisions a bit slicker E.g. if you already have 2.3456326 in the display and you want to do 12 ÷ 2.3456326, then you can just press [÷] [12] [=] [1/x] [=], which does the division the wrong way up and then flips it the right way up.
2) Analysing decimals to see if they might be rational, e.g. if the display is 0.1428571 and you press [1/x] [=] you'll get 7, meaning it was 1/7 before.

The Acid Test:

LEARN all the BITS AND BOBS on this page then TURN OVER and see how much you can remember.

1) x and y are measured as 2.32 m and 0.45 m to the nearest 0.01 m.
   a) Find the upper and lower bounds of x and y.
   b) If z = x + 1/y, find the min and max possible values of z.

Careful here — the biggest input values don't always give the biggest result.

Ratios

The whole grisly subject of RATIOS gets a whole lot easier when you do this:

Treat RATIOS like FRACTIONS

So for the RATIO 3:4, you'd treat it as the FRACTION 3/4, which is 0.75 as a DECIMAL.

What the fraction form of the ratio actually means

Suppose in a class there's girls and boys in the ratio 3 : 4.
This means there's 3/4 as many girls as boys.
So if there were 20 boys, there would be 3/4 × 20 = 15 girls.
You've got to be careful — it doesn't mean 3/4 of the people in the class are girls.

Reducing Ratios to their simplest form

You reduce ratios just like you'd reduce fractions to their simplest form.

For the ratio 15 : 18, both numbers have a factor of 3, so divide them by 3 — that gives 5 : 6. We can't reduce this any further. So the simplest form of 15 : 18 is 5 : 6.

Treat them just like fractions — use your calculator if you can

Now this is really sneaky. If you stick in a fraction using the $a^b/_c$ button, your calculator automatically cancels it down when you press =.
So for the ratio 8 : 12, just press 8 $a^b/_c$ 12 = , and you'll get the reduced fraction 2/3. Now you just change it back to ratio form ie. 2 : 3. Ace.

The More Awkward Cases:

1) The $a^b/_c$ button will only accept whole numbers

So IF THE RATIO IS AWKWARD (like "2.4 : 3.6" or "1¼ : 3½") then you must: MULTIPLY BOTH SIDES by the SAME NUMBER until they are both WHOLE NUMBERS and then you can use the $a^b/_c$ button as before to simplify them down.
e.g. with "1¼ : 3½", × both sides by 4 gives "5 : 14" (Try $a^b/_c$, but it won't cancel further)

2) If the ratio is MIXED UNITS

then you must CONVERT BOTH SIDES into the SMALLER UNITS using the relevant CONVERSION FACTOR (see P.24), and then carry on as normal.
e.g. "24mm : 7.2cm" (× 7.2cm by 10) ⇒ 24mm : 72mm = 1 : 3 (using $a^b/_c$)

3) To reduce a ratio to the form 1 : n or n : 1 (n can be any number)

Simply DIVIDE BOTH SIDES BY THE SMALLEST SIDE.
e.g. take "3 : 56" — dividing both sides by 3 gives: 1 : 18.7 (56÷3) (i.e. 1 : n)
This form is often the most useful, since it shows the ratio very clearly.

Ratios

Using The Formula Triangle in Ratio Questions

"Mortar is made from sand and cement in the ratio 7:2.
If 9 buckets of sand are used, how much cement is needed?"

This is a fairly common type of Exam question and it's pretty tricky for most people — but once you start using the formula triangle method, it's all a bit of a breeze really.

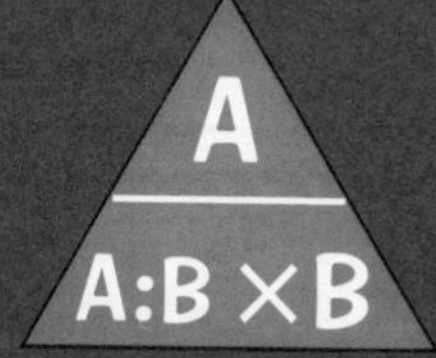

This is the basic FORMULA TRIANGLE for RATIOS, but NOTE:

1) THE RATIO MUST BE THE RIGHT WAY ROUND, with the FIRST NUMBER IN THE RATIO relating to the item ON TOP in the triangle.

2) You'll always need to CONVERT THE RATIO into its EQUIVALENT FRACTION or Decimal to work out the answer.

The formula triangle for the mortar question is shown below and the trick is to replace the RATIO 7:2 by its EQUIVALENT FRACTION: 7/2, or 3.5 as a decimal (7÷2)

So, covering up cement in the triangle, gives us "cement = sand / (7:2)"
i.e. "9 / 3.5" = 9 ÷ 3.5 = 2.57 or about 2½ buckets of cement.

Proportional Division

In a proportional division question a TOTAL AMOUNT is to be split in a certain ratio.

For example: "£9100 is to be split in the ratio 2:4:7. Find the 3 amounts".

The key word here is PARTS — concentrate on "parts" and it all becomes quite painless:

1) ADD UP THE PARTS:

The ratio 2:4:7 means there will be a total of 13 parts i.e. 2+4+7 = 13 PARTS

2) FIND THE AMOUNT FOR ONE "PART"

Just divide the total amount by the number of parts: £9100 ÷ 13 = £700 (= 1 PART)

3) HENCE FIND THE THREE AMOUNTS:

2 parts = 2×700 = £1400, 4 parts = 4×700 = £2800, 7 parts = £4900

The Acid Test:

LEARN the 6 RULES for SIMPLIFYING, the FORMULA TRIANGLE for Ratios (plus 2 points), and the 3 Steps for PROPORTIONAL DIVISION.

Now turn over and write down what you've learned. Try again until you can do it.

1) Simplify: a) 25:35 b) 3.4 : 5.1 c) 2¼ : 3¾
2) Porridge and ice-cream are mixed in the ratio 7:4 . How much porridge should go with 10 bowls of ice-cream?
3) Divide £8400 in the ratio 5:3:4

Fractions

This page shows you how to cope with fraction calculations without your beloved calculator.

By Hand

1) Multiplying — easy

Multiply top and bottom separately:

$$3/5 \times 4/7 = {}^{3\times4}/_{5\times7} = 12/35$$

2) Dividing — quite easy

Turn the 2nd fraction UPSIDE DOWN and then multiply:

$$3/4 \div 1/3 = 3/4 \times 3/1 = {}^{3\times3}/_{4\times1} = 9/4$$

3) Adding, subtracting — fraught

Add or subtract TOP LINES ONLY but only if the bottom numbers are the same. (If they're not, you have to make them the same – see below.)

$$2/6 + 1/6 = 3/6$$

$$5/7 - 3/7 = 2/7$$

4) Cancelling down — easy

Divide top and bottom by the same number, till they won't go any further:

$$18/24 \overset{\div 3}{=} 6/8 \overset{\div 2}{=} 3/4$$

(÷3 top and bottom, then ÷2 top and bottom)

5) Finding a fraction of something — just multiply

Multiply the 'something' by the TOP of the fraction, then divide it by the BOTTOM:

$$\frac{9}{20} \text{ of } £360 = \{(9) \times £360\} \div (20) = \frac{£3240}{20} = £162$$

or: $$\frac{9}{20} \text{ of } £360 = \frac{9}{1} \times £360 \times \frac{1}{20} = £162$$

6) Equalising the Denominator

This comes in handy for ordering fractions by size, and for adding or subtracting fractions. You need to find a common multiple of all the denominators:

Example: Put these fractions in ascending order of size: $8/3$, $6/4$, $12/5$

Lowest Common Multiple of 3, 4 and 5 is 60 so put all the fractions over 60... $\Longrightarrow$

$$8/3 = 8/3 \times 20/20 = 160/60$$
$$6/4 = 6/4 \times 15/15 = 90/60$$
$$12/5 = 12/5 \times 12/12 = 144/60$$

So the correct order is $90/60$, $144/60$, $160/60$, i.e. $6/4$, $12/5$, $8/3$

The Acid Test

Try all of the following without a calculator.

1) a) $3/8 \times 5/12$ b) $4/5 \div 7/8$ c) $3/4 + 2/5$ d) $2/5 - 3/8$ e) $4\frac{1}{9} + 2\frac{2}{27}$

2) a) Find $2/5$ of 550. b) What's $7/8$ of £2?

Fractions

Converting fractions to decimals is dead easy — remember that "/" means " ÷ " so all you have to do is divide, e.g. ¼ means 1 ÷ 4 = 0.25. Converting the other way is more fiddly. There's two methods, depending on whether you've got a terminating or recurring decimal...

Recurring or Terminating...

Recurring decimals have a pattern of numbers which repeats forever, e.g. $\frac{1}{3}$ is the decimal 0.333333... Terminating decimals are finite, e.g. $\frac{1}{20}$ is the decimal 0.05.

The denominator (bottom number) of a fraction, tells you if it'll be a recurring or terminating decimal when you convert it. Fractions where the denominator has prime factors of only 2 or 5 will give terminating decimals. All other fractions will give recurring decimals.

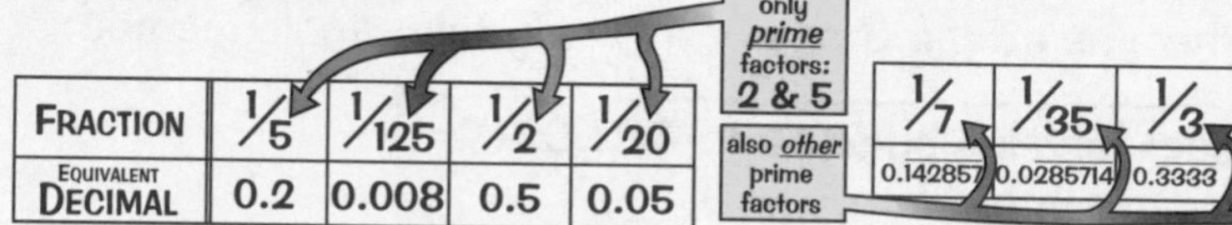

FRACTION	$\frac{1}{5}$	$\frac{1}{125}$	$\frac{1}{2}$	$\frac{1}{20}$
EQUIVALENT DECIMAL	0.2	0.008	0.5	0.05

$\frac{1}{7}$	$\frac{1}{35}$	$\frac{1}{3}$	$\frac{1}{6}$
0.142857	0.0285714	0.3333	0.16666

For prime factors see p.19

Terminating Decimals into Fractions

You should already know this easy method for converting terminating decimals — just divide by a power of 10 which depends on the number of digits after the decimal point.

E.g. 0.6 = $\frac{6}{10}$, 0.78 = $\frac{78}{100}$, 0.908 = $\frac{908}{1000}$, etc.

Recurring Decimals into Fractions

There's two ways to do it: 1) by UNDERSTANDING 2) by just LEARNING THE RESULT.

The Understanding Method:

1) Find the length of the repeating sequence and multiply by 10, 100, 1000, 10 000 or whatever to move it all up past the decimal point by one full repeated lump:

E.g. 0.234234234... × 1000 = 234.234234..

2) Subtract the original number, r, from the new one (which in this case is 1000r)
i.e. 1000r − r = 234.234234... − 0.234234... giving: 999r = 234

3) Then just DIVIDE to leave r: $r = \frac{234}{999}$, and cancel if possible: $r = \frac{26}{111}$

The "Just Learning The Result" Method:

The fraction always has the repeating unit on the top and the same number of nines on the bottom — easy as that. Look at these and marvel at the elegant simplicity of it.

0.4444444 = 4/9 0.34343434 = 34/99
0.124124124 = 124/999 0.14561456 = 1456/9999

Always check if it will CANCEL DOWN of course, e.g. 0.363636.... = 36/99 = 4/11.

The Acid Test:

LEARN how to tell whether a fraction will be a terminating or recurring decimal, and all the methods above. Then turn over and write it all down.

1) Express 0.142857142857.... as a fraction. 2) Convert 0.035 into a fraction, and cancel it down.

Percentages

You shouldn't have any trouble with most percentage questions, especially types 1 and 2. However watch out for type 3 questions and make sure you know the proper method for doing them. "Percentage change" can also catch you out if you don't watch all the details – using the ORIGINAL value for example.

Type 1

"Find x% of y" — e.g. Find 15% of £46 ⇒ 0.15 ×46 = £6.90

Type 2

"Express x as a percentage of y"
e.g. Give 40p as a percentage of £3.34 ⇒ (40 ÷ 334) × 100 = 12%

Type 3

— IDENTIFIED BY NOT GIVING THE "ORIGINAL VALUE"

These are the type most people get wrong – but only because they don't recognise them as a type 3 and don't apply this simple method:

Example: A house increases in value by 20% to £72,000. Find what it was worth before the rise.

Method:

£72,000 = 120%

÷120

£600 = 1%

×100

£60,000 = 100%

So the original price was £60,000

An INCREASE of 20% means that £72,000 represents 120% of the original value. If it was a DROP of 20%, then we would put "£72,000 = 80%" instead, and then divide by 80 on the LHS, instead of 120.

Always set them out exactly like this example. The trickiest bit is deciding the top % figure on the RHS — the 2nd and 3rd rows are always 1% and 100%

Percentage Change

It is common to give a change in value as a percentage.
This is the formula for doing so — LEARN IT, AND USE IT:

$$\text{PERCENTAGE "CHANGE"} = \frac{\text{"CHANGE"}}{\text{ORIGINAL}} \times 100$$

By "change", we could mean all sorts of things such as: "Profit", "loss", "appreciation", "depreciation", "increase", "decrease", "error", "discount", etc. For example,

$$\text{percentage "profit"} = \frac{\text{"profit"}}{\text{original}} \times 100$$

Note the great importance of using the ORIGINAL VALUE in this formula.

The Acid Test:

LEARN The details for TYPE 3 QUESTIONS and PERCENTAGE CHANGE, then turn over and write it all down.

1) A trader buys watches for £5 and sells them for £7. Find his profit as a percentage.
2) A car depreciates by 30% to £14,350. What was it worth before?
3) Find the percentage error in rounding 3.452 to 3.5. Give your answer to 2 DP.

Regular Polygons

A polygon is a many-sided shape. A regular polygon is one where all the sides and angles are the same. The regular polygons are a never-ending series of shapes with some fancy features. They're very easy to learn. Here are the first few but they don't stop – you can have one with 12 sides or 25, etc.

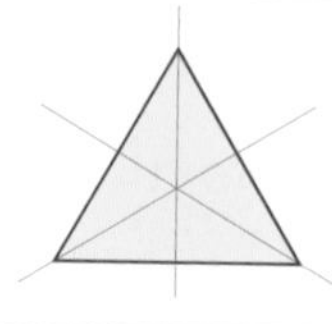

EQUILATERAL TRIANGLE
3 sides
3 lines of symmetry
Rot^nl symm. order 3

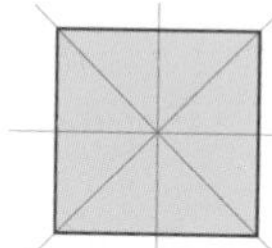

SQUARE
4 sides
4 lines of symmetry
Rot^nl symm. order 4

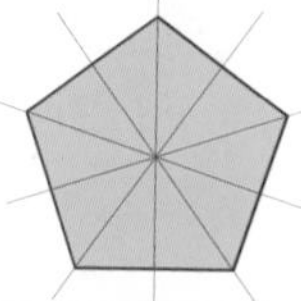

REGULAR PENTAGON
5 sides
5 lines of symmetry
Rot^nl symm. order 5

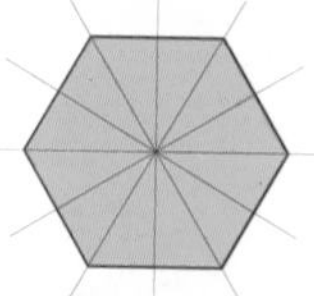

REGULAR HEXAGON
6 sides
6 lines of symmetry
Rot^nl symm. order 6

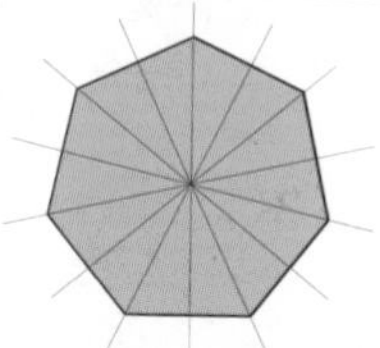

REGULAR HEPTAGON
7 sides
7 lines of symmetry
Rot^nl symm. order 7
(A 50p piece is like a heptagon)

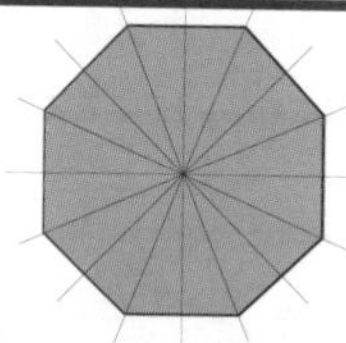

REGULAR OCTAGON
8 sides
8 lines of symmetry
Rot^nl symm. order 8

Interior And Exterior Angles

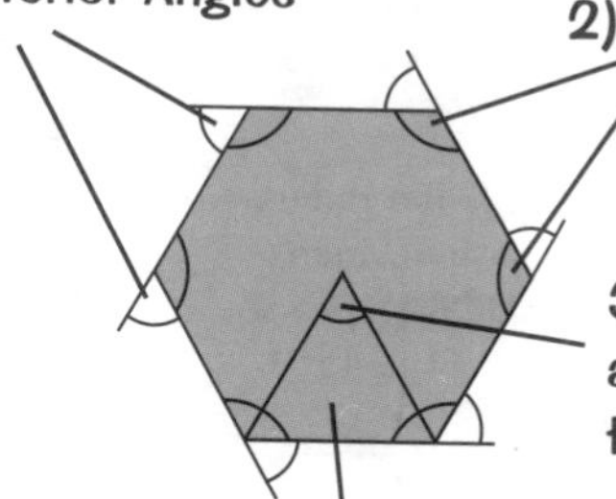

There are 4 formulas to learn:

$$\text{EXTERIOR ANGLE} = \frac{360^\circ}{n}$$

$$\text{INTERIOR ANGLE} = 180^\circ - \text{EXTERIOR ANGLE}$$

$$\text{SUM OF EXTERIOR ANGLES} = 360^\circ$$

$$\text{SUM OF INTERIOR ANGLES} = (n-2) \times 180^\circ$$

(n is the number of sides)

Note — the two SUM formulas above work for any polygons, not just regular ones,

You also need to know the next two, but I'm not drawing them for you. Learn their names:

REGULAR NONAGON
9 sides, etc. etc.

REGULAR DECAGON
10 sides, etc. etc.

REGULAR POLYGONS HAVE LOADS OF SYMMETRY

1) The pentagon shown here has only 3 different angles in the whole diagram.
2) This is typical of regular polygons. They display an amazing amount of symmetry.
3) With a regular polygon, if two angles look the same, they will be. That's not a rule you should normally apply in geometry, and anyway you'll need to prove they're equal.

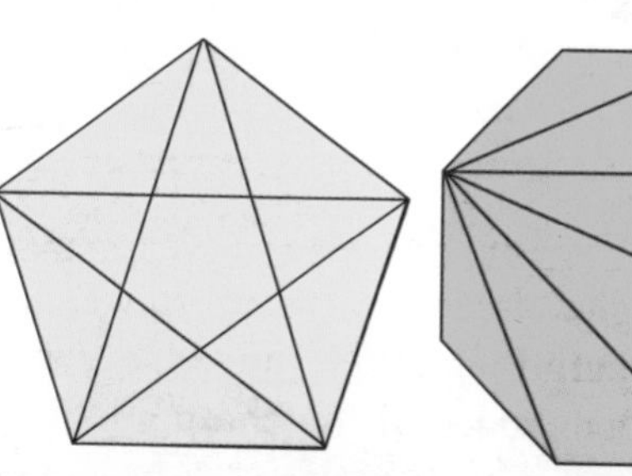

The Acid Test:

LEARN THIS PAGE.
Then turn over and write down everything.

1) Work out the two key angles for a Pentagon
2) And for a 12-sided Regular Polygon.
3) A regular polygon has an interior angle of 156°. How many sides does it have?
4) Work out ALL the angles in the pentagon and octagon shown above — and marvel at the symmetry.

Circles

Circle Formulas — Areas and Lengths

AREA AND CIRCUMFERENCE

You should know these already, but here's a quick reminder...

AREA of CIRCLE = $\pi \times (\text{radius})^2$

$$A = \pi \times r^2$$

$\pi = 3.141592....$
$= 3.14$ (approx)

e.g. if the radius is 4cm, then
$A = 3.14 \times (4 \times 4)$
$= 50.24\text{cm}^2$

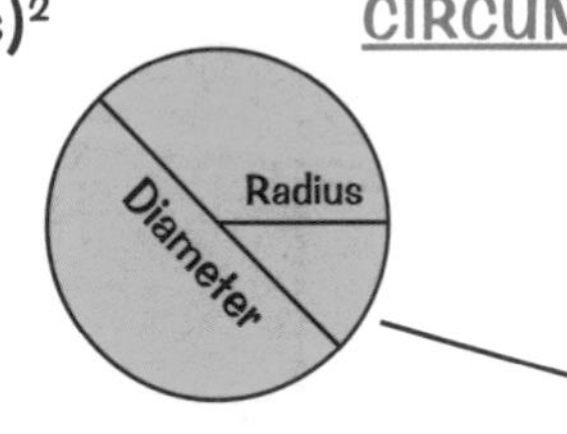

CIRCUMFERENCE = $\pi \times$ diameter

$$C = \pi \times D$$

Circumference = distance round the outside of the circle

SECTORS AND ARCS

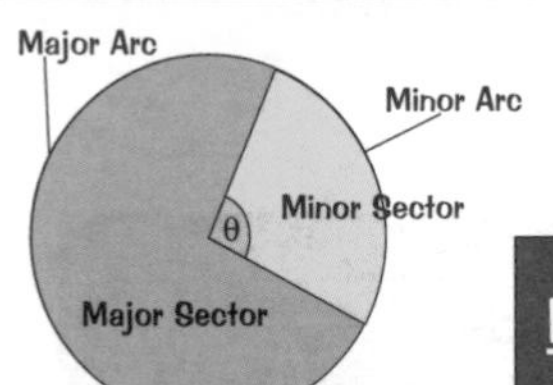

$$\text{Area of Sector} = \frac{\theta}{360} \times \text{Area of full Circle}$$

(Pretty obvious really isn't it?)

$$\text{Length of Arc} = \frac{\theta}{360} \times \text{Circumference of full Circle}$$

(Obvious again, no?)

AREA OF A SEGMENT

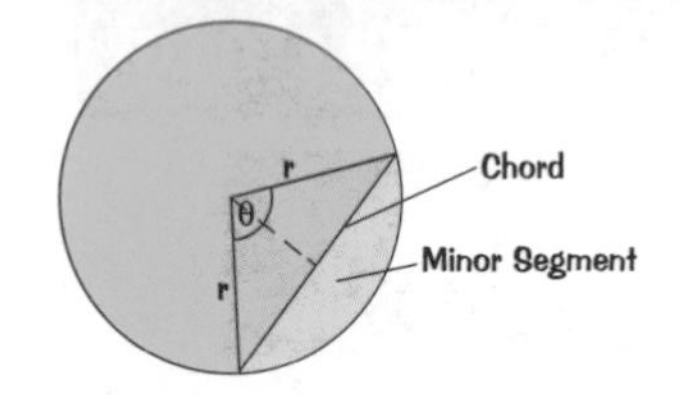

FINDING THE AREA OF A SEGMENT is a slightly involved business but worth learning:

1) Find the area of the sector as above.
2) Find the area of the triangle. You can use $A = \frac{1}{2} ab \sin C$, or in this case, $A = \frac{1}{2} r^2 \sin \theta$.
3) Then just subtract the area of the triangle from the area of the sector.

Circle Geometry — Six Rules

1) ANGLE IN A SEMICIRCLE = 90°

A triangle drawn from the two ends of a diameter will always make an angle of 90° where it meets the edge of the circle.

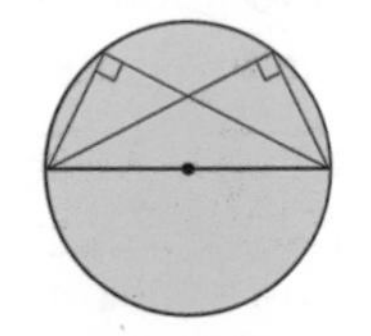

2) ANGLES IN THE SAME SEGMENT ARE EQUAL

All triangles drawn from a chord will have the same angle where they touch the circle. Also, the two angles on opposite sides of the chord add up to 180°.

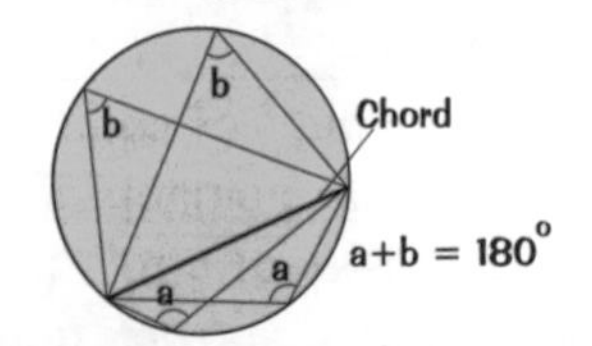

3) ANGLE AT THE CENTRE IS TWICE THE ANGLE AT THE EDGE

The angle subtended at (posh way of saying "angle made at") the centre of a circle is exactly double the angle subtended at the edge of the circle from the same two points (two ends of the same chord).

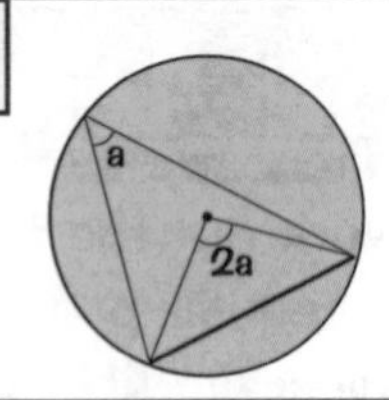

Circles

4) OPPOSITE ANGLES OF A CYCLIC QUADRILATERAL ADD UP TO 180°

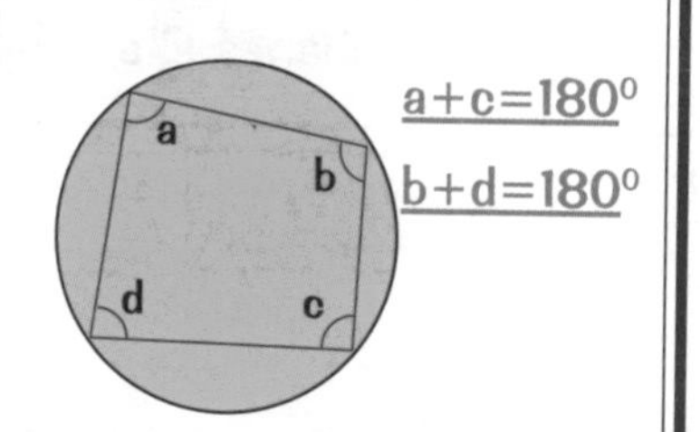

A cyclic quadrilateral is a 4-sided shape with every corner touching the circle. Both pairs of opposite angles add up to 180°.

5) ANGLE IN OPPOSITE SEGMENT IS EQUAL

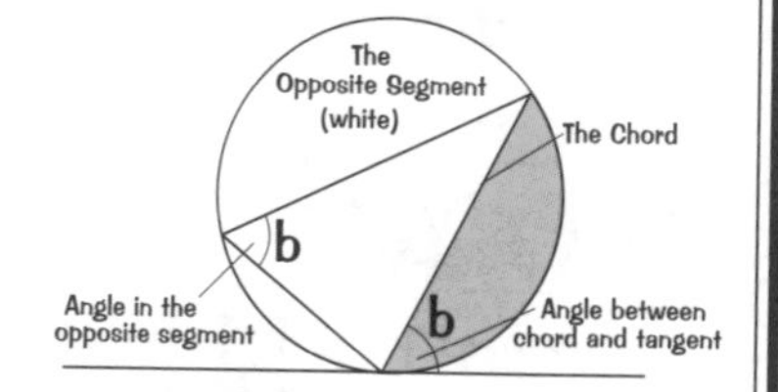

This is quite a tricky one to remember. If you draw a tangent and a chord that meet, then the angle between them is always equal to "the angle in the opposite segment" (i.e. the angle made at the edge of the circle by two lines drawn from the chord).

6) A CHORD BISECTOR IS A DIAMETER

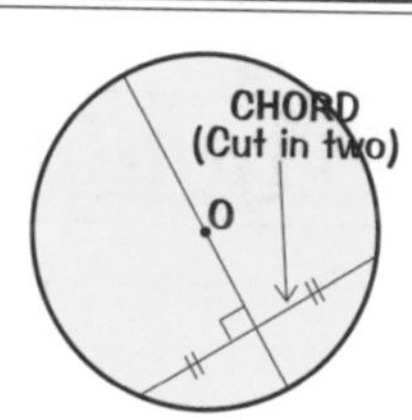

A CHORD is any line drawn across a circle. And no matter where you draw a chord, the perpendicular line (at 90°) that cuts it exactly in half, will go through the centre of the circle and so will be a DIAMETER.

Example

"Find all the angles in this diagram."

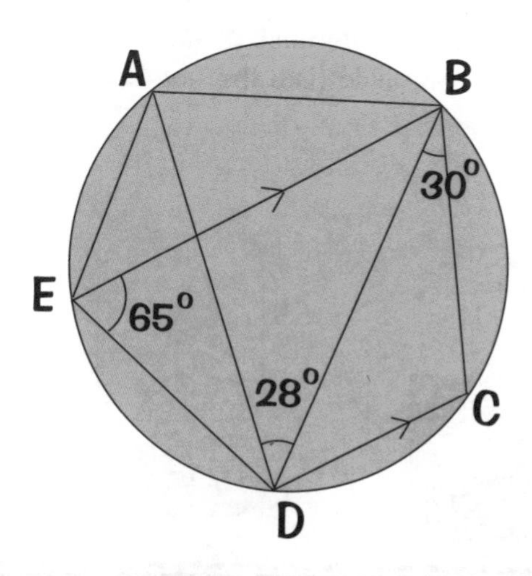

1) PARALLEL LINES — there are actually 4 different lines crossing the 2 parallel ones, but the most useful one is ED which tells us that EDC is 115°

2) ANGLE IN SAME SEGMENT — there are potentially eight different chords where this rule could apply, but some are more useful than others:
   EAD = EBD, ADB = AEB (so AEB = 28°)
   ABE = ADE, DAB = DEB (so DAB = 65°)

3) OPPOSITE ANGLES OF A CYCLIC QUADRILATERAL
   — looking at BEDC gives:
   BCD = 180 – DEB = 180 – 65 = 115°
   — looking at ABDE gives:
   ABD = 180 – AED = 180 – (28 + 65) = 87°

4) ANGLES IN A TRIANGLE ADD UP TO 180° — this, the simplest of all the rules, will now find all the other angles for you.

The Acid Test:

LEARN all the circle formulas and the Six Geometry Rules Then turn over and write them all down.

1) Practise the above Example till you understand every step and can do it easily without help.

2) Find all the angles in the diagram to the right, illustrating the 3-letter notation (ODC = 48°, etc.).

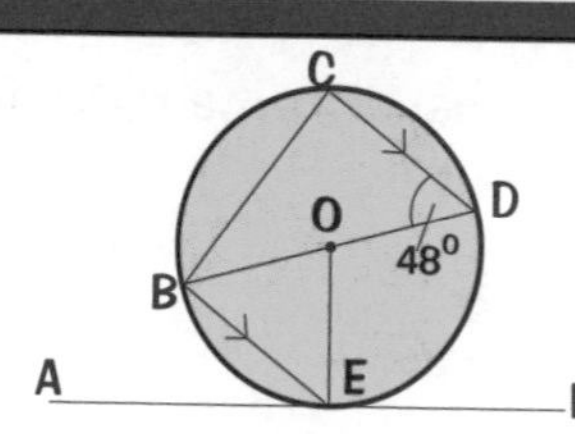

Volumes

You should know how to calculate the volume of simple shapes like cuboids.
For harder shapes like spheres and cones all you can do is learn the formula...

1) Sphere

Volume of sphere = $\frac{4}{3}\pi r^3$

EXAMPLE: The moon has a radius of 1700km, find its volume.

Ans: $V = \frac{4}{3}\pi r^3 = (4/3) \times 3.14 \times 1700^3 = 2.1 \times 10^{10}$ km³ (A lot of cheese)

2) Prisms

Finding the volumes of prisms was covered on page 35.

But just to remind you: **Volume of prism = Cross-sectional Area × length**

$$V = A \times l$$

3) Pyramids and Cones

A pyramid is any shape that goes up to a point at the top. Its base can be any shape at all.
If the base is a circle then it's called a cone (rather than a circular pyramid).

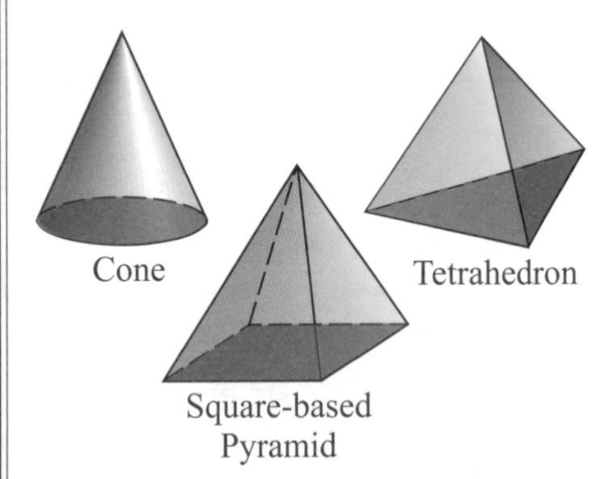
Cone Tetrahedron Square-based Pyramid

Volume of Pyramid = $\frac{1}{3}$ x Base Area x Height

Volume of Cone = $\frac{1}{3} \times \pi r^2 \times$ Height

This surprisingly simple formula is true for any pyramid or cone, whether it goes up "vertically" (like the three shown here) or off to one side (like the one below).

4) A Frustum is Part of a Cone

A frustum of a cone is what's left when the top part of a cone is cut off parallel to its circular base

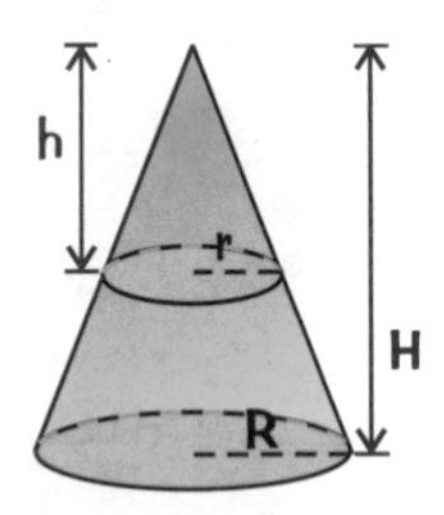

VOLUME OF FRUSTUM = VOLUME OF THE ORIGINAL CONE − VOLUME OF THE REMOVED CONE

$$= \frac{1}{3}\pi R^2 H - \frac{1}{3}\pi r^2 h$$

The Acid Test:

LEARN this page. Then turn over and try to write it all down. Keep trying until you can do it.

1) Name the shape to the right and find its volume:

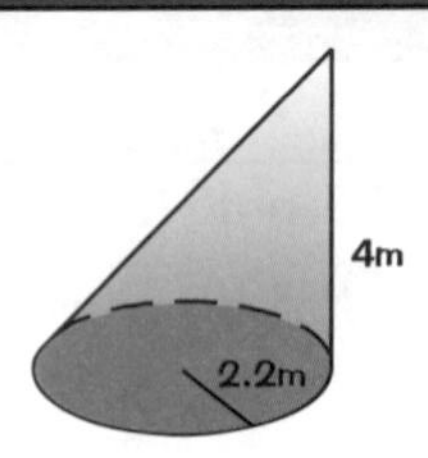

2) A ping pong ball has a diameter of 4cm. A tennis ball has a diameter of 7cm. Find the volume of both balls.

Surface Area and Projections

As with volume, you should already know how to find the surface area of many simple shapes such as cuboids and prisms. But for other shapes, you just have to learn the formulas...

Surface Area and Nets

Remember, the SURFACE AREA of a 3D object is just the total area of all the outer surfaces added together. Prisms, cubes, cuboids and pyramids were covered on p.34.

The nets for spheres, cones and cylinders are difficult to draw, so make sure you learn the formulas...

CYLINDERS:
Surface area $= 2\pi rh + 2\pi r^2$

Note that the length of the rectangle is equal to the circumference of the circular ends.

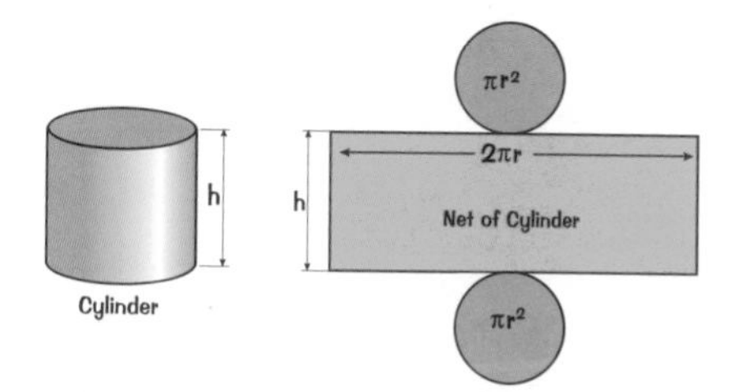

SPHERES:
Surface area $= 4\pi r^2$

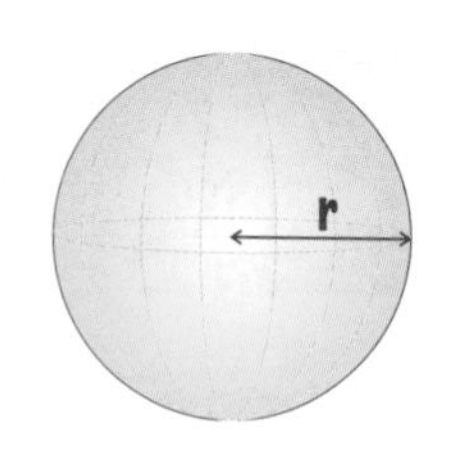

CONES:
Surface area $= \pi rl + \pi r^2$

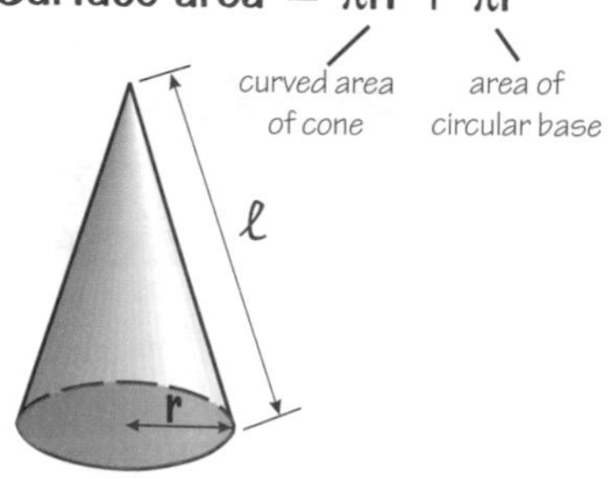

Projections show Different Viewpoints

A 'projection' shows the relative size and shape of an object from either the front, side or back — they're usually known as 'elevations'. A 'plan' shows the view from above. They're always drawn to scale.

FRONT Elevation — the view you'd see if you looked from directly in front:

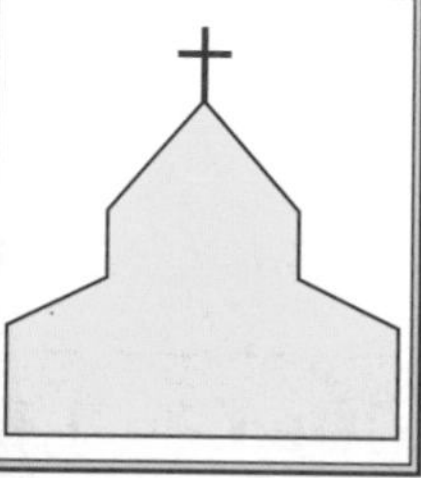

SIDE Elevation — the view you'd see if you looked from directly to one side:

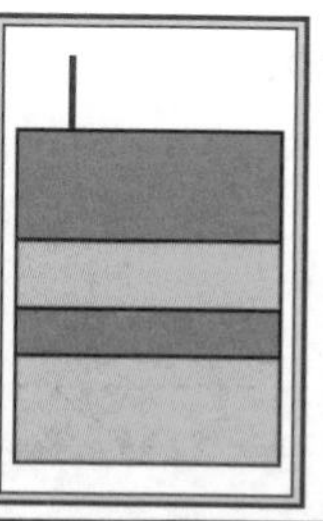

PLAN — the view you'd see if you looked from directly above:

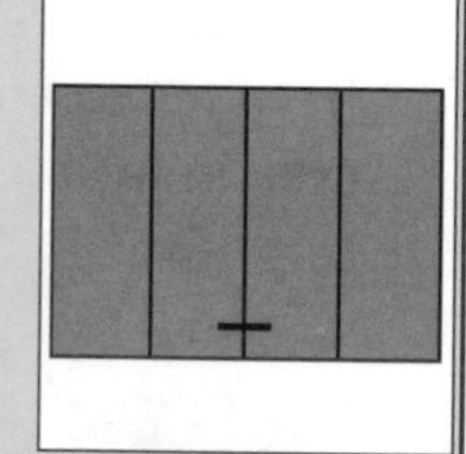

If they're feeling really mean (and they often are), you might get a question on:

This one's a bit trickier, so you might want to spend a little longer practising it — just to get your head round it.

ISOMETRIC Projection — this is where the shape is drawn (again, to scale) from a view at equal angles to all three axes (x, y and z). Or more simply, it's a drawing like this:

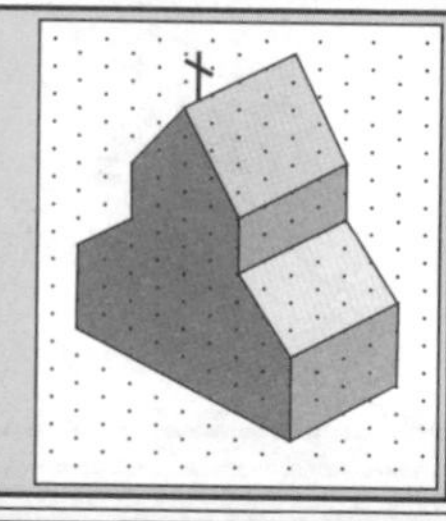

The Acid Test:

LEARN this page. Then turn over and try to write it all down.

1) Calculate the surface area of the two ping pong balls from the previous page.
2) Work out the surface area of a cylindrical drink can of height 12.5cm and diameter 7.2cm.
3) Draw plan, front and side elevations and an isometric projection of your house.

Length, Area and Volume

Identifying Formulas Just by Looking at Them

This isn't as bad as it sounds, since we're only talking about the formulas for 3 things:

LENGTH, AREA and VOLUME

The rules are as simple as this:

AREA FORMULAS always have LENGTHS MULTIPLIED IN PAIRS

VOLUME FORMULAS always have LENGTHS MULTIPLIED IN GROUPS OF THREE

LENGTH FORMULAS (such as perimeter) always have LENGTHS OCCURRING SINGLY

In formulas of course, lengths are represented by letters, so when you look at a formula you're looking for: groups of letters multiplied together in ones, twos or threes.

But remember, π is not a length.

Examples:

r^2 means $r \times r$, don't forget (see p.28)

$4\pi r^2 + 6d^2$	(area)	$Lwh + 6r^2L$	(volume)
$4\pi r + 15L$	(length)	$6hp + \pi r^2 + 7h^2$	(area)
$3p(2b + a)$	(area)	$3\pi h(L^2 + 4P^2)$	(volume)

Watch out for tricky ones with brackets — you should multiply out the brackets first (see p.29)

Converting Area and Volume Measurements

$$1m^2 = 100cm \times 100cm = 10,000cm^2$$

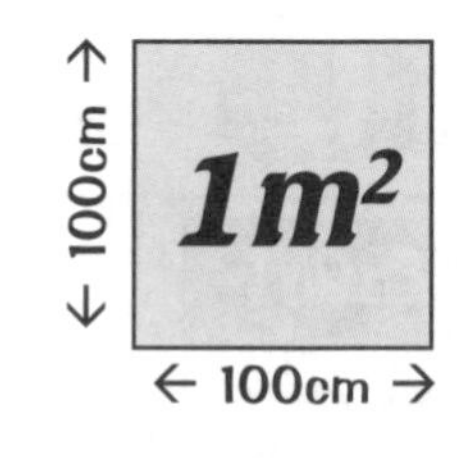

1) To change area measurements from m^2 to cm^2 multiply the area in m^2 by 10,000 (e.g. $3m^2 = 30,000cm^2$).

2) To change area measurements from cm^2 to m^2 divide the area in cm^2 by 10,000 (e.g. $45,000cm^2 = 4.5m^2$).

$$1m^3 = 100cm \times 100cm \times 100cm = 1,000,000cm^3$$

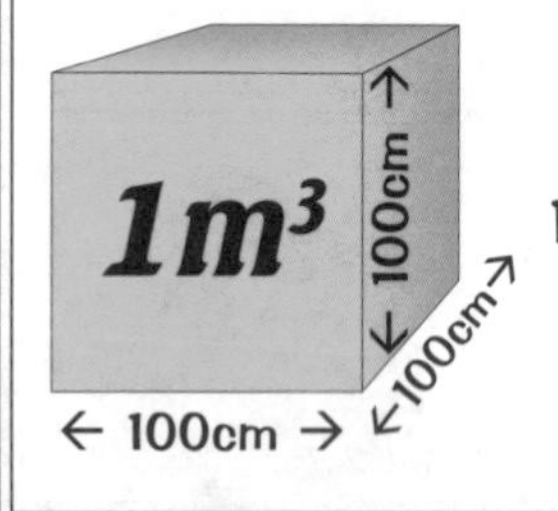

1) To change volume measurements from m^3 to cm^3 multiply the volume in m^3 by 1,000,000 (e.g. $3m^3 = 3,000,000cm^3$).

2) To change volume measurements from cm^3 to m^3 divide the volume in cm^3 by 1,000,000 (e.g. $4,500,000cm^3 = 4.5m^3$).

The Acid Test:

LEARN the Rules for Identifying Formulas.
Turn over and write it all down.

1) Identify each of these expressions as an area, volume, or perimeter:

πr^2 $\quad Lwh$ $\quad \pi d$ $\quad ½bh$ $\quad 2bh + 4l\pi$ $\quad 4r^2h + 3\pi d^3$ $\quad 2\pi r(3L + 5T)$

2) Convert these area measurements: a) $23\ m^2 \rightarrow cm^2$ b) $34,500\ cm^2 \rightarrow m^2$

3) Convert these volume measurements: a) $5.2m^3 \rightarrow cm^3$ b) $100,000\ cm^3 \rightarrow m^3$

D/T Graphs and V/T Graphs

Distance-time graphs and Velocity-time graphs are so common in Exams that they deserve a page all to themselves just to make sure you know all the vital details about them.
The best thing about them is that they don't vary much and they're always easy.

1) Distance-Time Graphs

Just remember these 3 important points:

1) At any point, GRADIENT = SPEED, but watch out for the UNITS.
2) The STEEPER the graph, the FASTER it's going.
3) FLAT SECTIONS are where it is STOPPED.

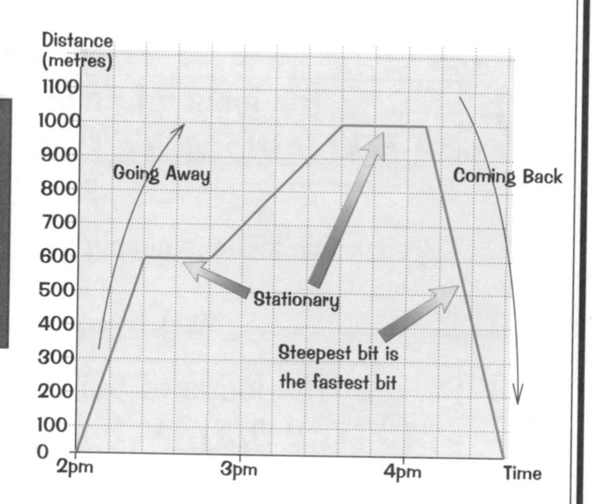

EXAMPLE: "What is the speed of the return section on the graph shown?"

Speed = gradient = 1000m/30mins = **33.33** m/min. But m/min are naff units so it's better to do it like this: 1km ÷ 0.5 hrs = 2 km/h

2) Velocity-Time Graphs

A Velocity-Time graph can look just the same as a Distance-Time graph but it means something completely different. The graph shown here is exactly the same shape as the one above, but the actual motions are completely different.

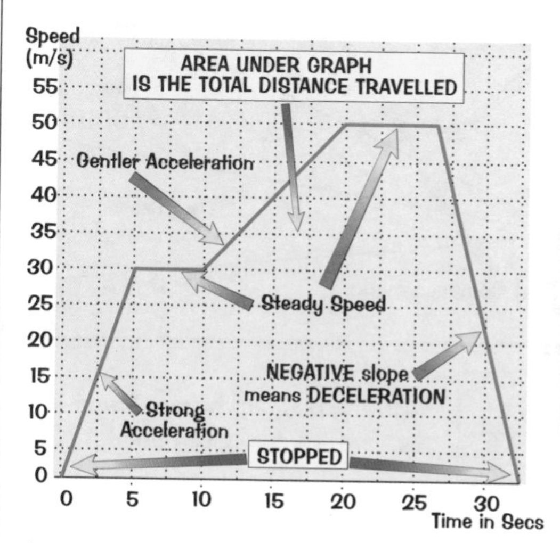

Remember these 4 important points:

1) At any point, GRADIENT = ACCELERATION, (The UNITS are m/s^2 don't forget).
2) NEGATIVE SLOPE is DECELERATION.
3) FLAT SECTIONS are STEADY SPEED.
4) AREA UNDER GRAPH = DISTANCE TRAVELLED

The D/T graph shows something moving away and then back again with steady speeds and long stops, rather like a donkey on Blackpool Beach. The V/T graph on the other hand shows something that sets off from rest, accelerates strongly, holds its speed, then accelerates again up to a maximum speed which it holds for a while and then comes to a dramatic halt at the end. MORE LIKE A FERRARI THAN A DONKEY!

The Acid Test:

LEARN the 7 IMPORTANT POINTS and the TWO DIAGRAMS then turn over and write them all down.

1) For the D/T graph shown above, work out the speed of the middle section in km/h.
2) For the V/T graph, work out the three different accelerations and the two steady speeds.

Loci and Construction

A LOCUS (another ridiculous maths word) is simply:

A LINE that shows all the points which fit in with a given rule

Make sure you learn how to do these **PROPERLY** using a ruler and compasses as shown on these two pages

1) The locus of points which are "A FIXED DISTANCE from a given POINT"

This locus is simply a CIRCLE.

Pair of Compasses

A given point

The LOCUS of points equidistant from it

2) The locus of points which are "A FIXED DISTANCE from a given LINE"

This locus is an OVAL SHAPE

It has straight sides (drawn with a ruler) and ends which are perfect semicircles (drawn with compasses).

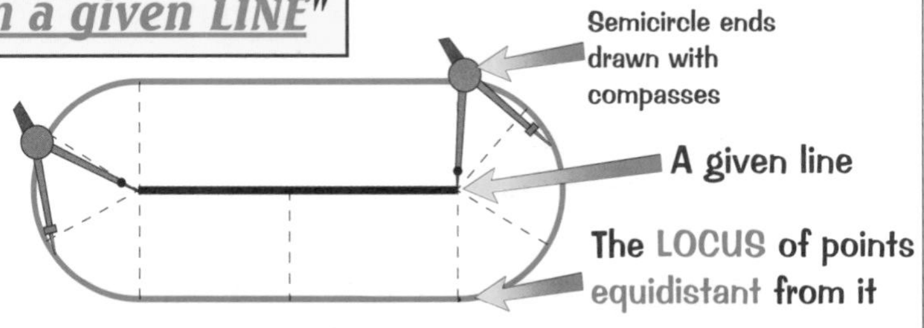

3) The locus of points which are "EQUIDISTANT from TWO GIVEN LINES"

1) Keep the compass setting THE SAME while you make all four marks.
2) Make sure you leave your compass marks showing.
3) You get two equal angles — i.e. this LOCUS is actually an ANGLE BISECTOR.

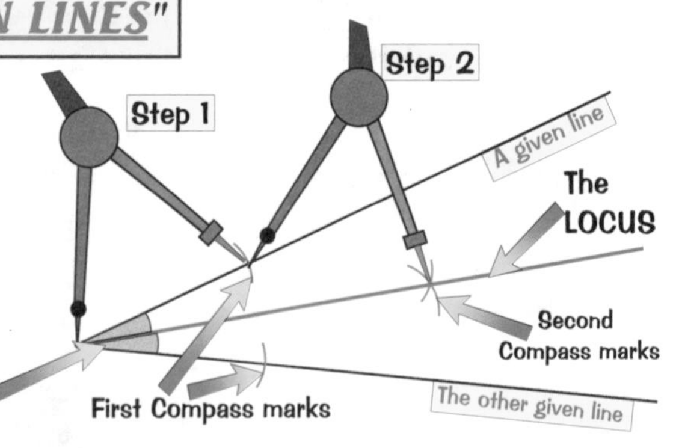

4) The locus of points which are "EQUIDISTANT from TWO GIVEN POINTS"

(In the diagram below, A and B are the two given points)

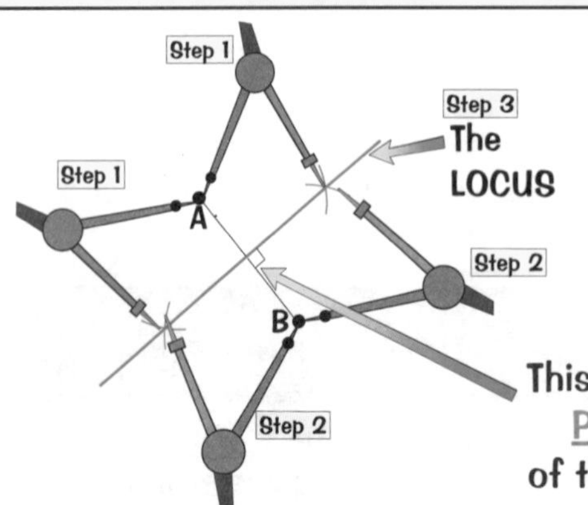

This LOCUS is all points which are the same distance from A as they are from B.

This time the locus is actually the PERPENDICULAR BISECTOR of the line joining the two points.

Loci and Construction

Constructing accurate 60° angles

1) They may well ask you to draw an accurate 60° angle.

2) One place they're needed is for drawing an equilateral triangle.

3) Make sure you follow the method shown in this diagram, and that you can do it entirely from memory.

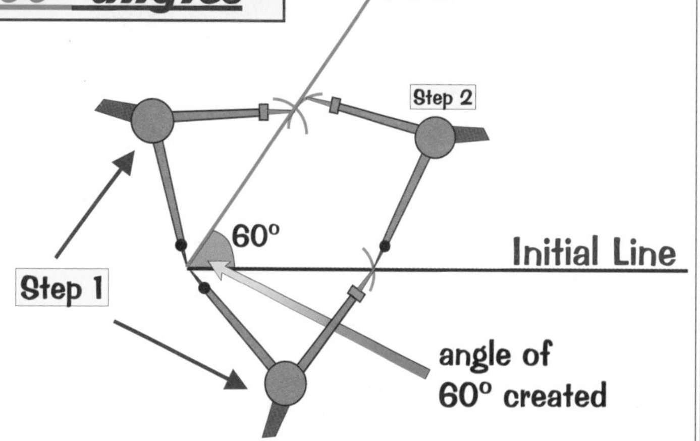

Constructing accurate 90° angles

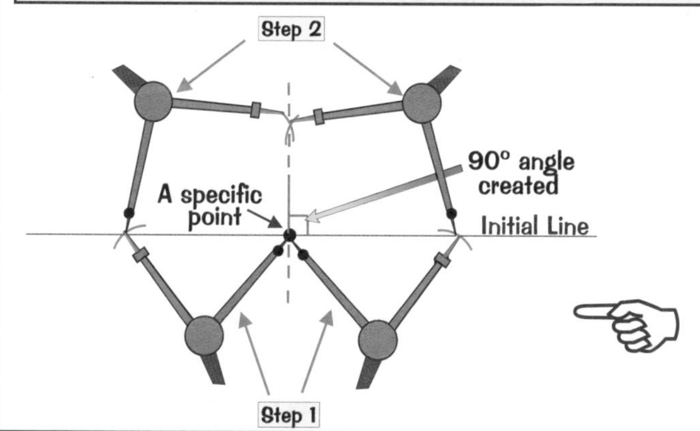

1) They might want you to draw an accurate 90° angle.
2) They won't accept it just done "by eye" or with a ruler — if you want the marks you've got to do it the proper way with compasses like I've shown you here.
3) Make sure you can follow the method shown in this diagram.

Drawing the Perpendicular from a Point to a Line

1) This is similar to the one above but not quite the same — make sure you can do both.

2) Again, they won't accept it just done "by eye" or with a ruler — you've got to do it the proper way with compasses.

3) Learn the diagram.

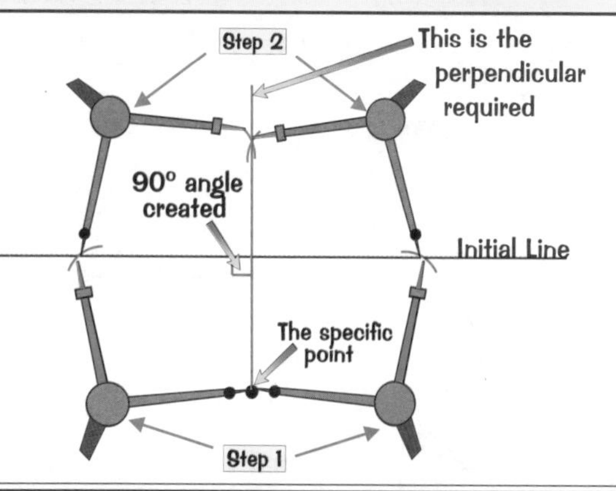

The Acid Test: LEARN EVERYTHING ON THESE TWO PAGES

Now cover up these two pages and draw an example of each of the four loci.
Also draw an equilateral triangle and a square with accurate 60° and 90° angles.
Also, draw a line and a point and construct the perpendicular from the point to the line.

Congruence and Similarity

Congruence is another ridiculous maths word which sounds really complicated when it's not:
If two shapes are congruent, they are simply the same — the same size and the same shape.
That's all it is. They can however be mirror images.

CONGRUENT — same size, same shape

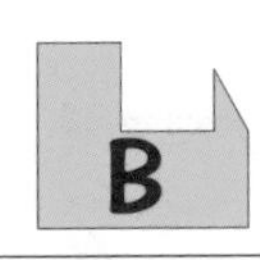

SIMILAR — same shape, different size

C D

Note that the angles are always unchanged

Congruent Triangles — are they or aren't they?

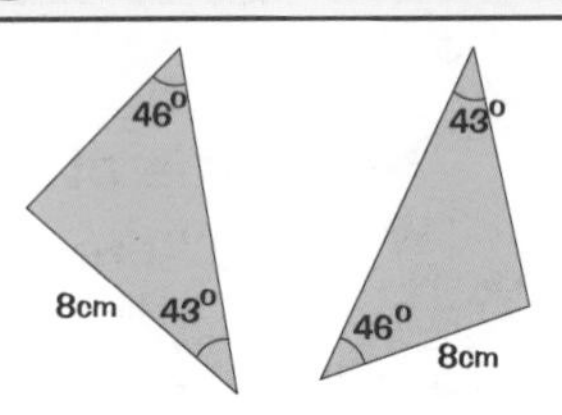

Probably the trickiest area of congruence is deciding whether two triangles, like the ones shown here, are CONGRUENT.
In other words, from the skimpy information given, are the two going to be the same or different.
There are THREE IMPORTANT STEPS:

1) The Golden Rule is definitely to DRAW THEM BOTH IN THE SAME ORIENTATION — only then can you compare them properly:

8cm 43° 46°   8cm 43° 46°

2) Don't jump to hasty conclusions — although the 8cm sides are clearly in different positions, it's always possible that both top sides are 8cm.
In this case we can work out that they're not because the angles are different (so they can't be isosceles).

3) Now see if any of these conditions are true. If ONE of the conditions holds, the triangles are congruent.

1) SSS three sides are the same
2) AAS two angles and a side match up
3) SAS two sides and the angle between them match up
4) RHS a right angle, the hypotenuse (longest side) and one other side all match up

For two triangles to be congruent, ONE OR MORE of these four conditions must hold.
(If none are true, then you have proof that the triangles aren't congruent).

Congruence and Transformations

WHEN A SHAPE IS TRANSLATED, ROTATED OR REFLECTED, THE IMAGE IS CONGRUENT TO THE ORIGINAL SHAPE. ENLARGEMENTS DON'T FOLLOW THIS RULE.

e.g.

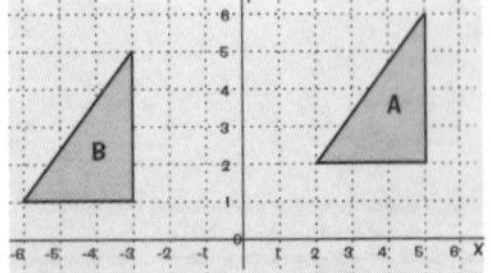

A to B is a translation of $\begin{pmatrix}-8\\-1\end{pmatrix}$.
The lengths and angles are unchanged, so A is congruent to B.

e.g.

A to B is an enlargement of scale factor 2, and centre (2, 6).

The angles are unchanged but not the lengths, so A is not congruent to B.

The Acid Test:

LEARN the definitions of similarity and congruence, the 3 steps for checking for congruent triangles, and the rule about transformations.

Then, when you think you know it, turn the page over and write it all down again, from memory, including the sketches and examples.

Similarity and Enlargements

4 Key Features

1) If the Scale Factor is bigger than 1 the shape gets bigger.

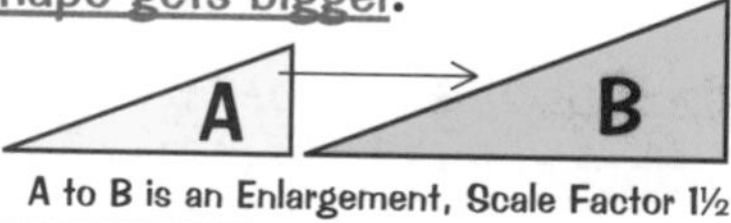

A to B is an Enlargement, Scale Factor 1½

2) If the Scale Factor is smaller than 1 (i.e. a fraction like ½) then the shape gets smaller. (Really this is a reduction, but you still call it an Enlargement, Scale Factor ½)

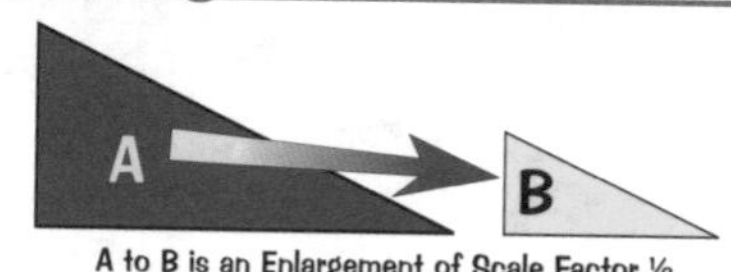

A to B is an Enlargement of Scale Factor ½

3) If the Scale Factor is NEGATIVE then the shape pops out the other side of the enlargement centre. If the scale factor is -1, it's exactly the same as a rotation of 180°

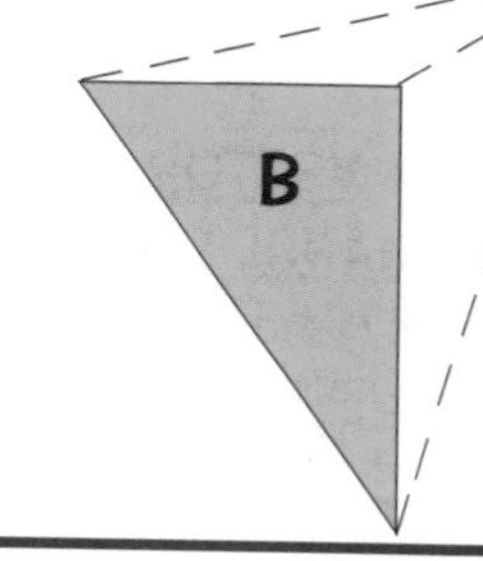

A to B is an enlargement of scale factor -2. B to A is an enlargement of scale factor -½.

4) The Scale Factor also tells you the relative distance of old points and new points from the Centre of Enlargement — this is very useful for drawing an enlargement, because you can use it to trace out the positions of the new points:

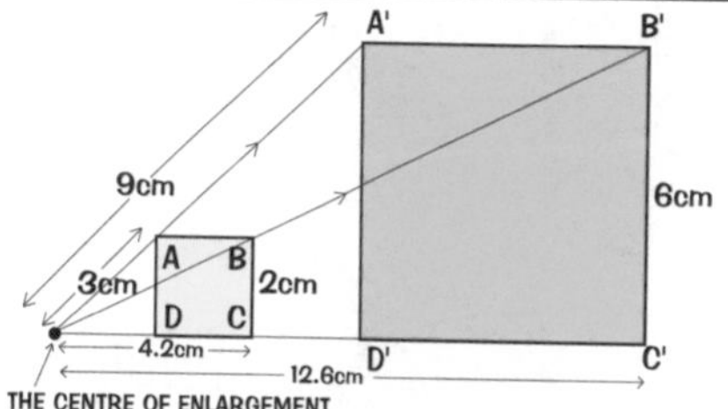

Areas and Volumes of Enlargements

Ho ho! This little joker catches everybody out. The increase in area and volume is BIGGER than the scale factor. For example, if the Scale Factor is 2, the lengths are twice as big, each area is 4 times as big, and the volume is 8 times as big. The rule is this:

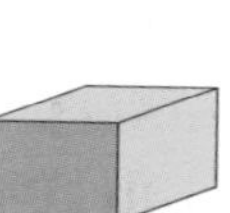

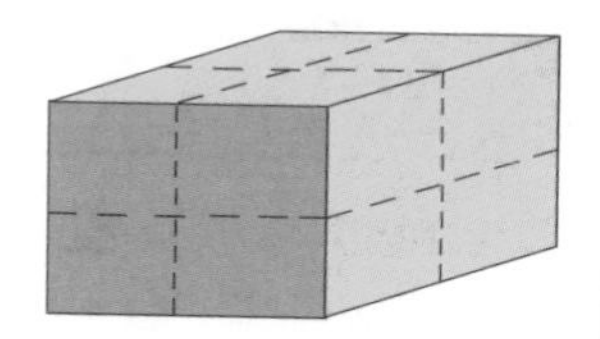

For a Scale Factor n:

The SIDES are	n times bigger
The AREAS are	n^2 times bigger
The VOLUMES are	n^3 times bigger

Simple... but VERY FORGETTABLE

These ratios can also be expressed in this form:

Lengths	$a : b$	e.g. 3 : 4
Areas	$a^2 : b^2$	e.g. 9 : 16
Volumes	$a^3 : b^3$	e.g. 27 : 64

EXAMPLE: 2 spheres have surface areas of 16m² and 25m². Find the ratio of their volumes.

ANS: 16 : 25 is the areas ratio which must be $a^2 : b^2$,
i.e. $a^2 : b^2 = 16 : 25$
and so $a : b = 4 : 5$
and so $a^3 : b^3 = 64 : 125$ The volumes ratio.

The Acid Test:

LEARN the 4 Key Features for Enlargements, plus the 3 Rules for Area and Volume Ratios. Then turn over and write them all down.

1) Draw the triangle A(2,1) B(5,2) C(4,4) and enlarge it by a scale factor of -1½, centre the origin. Label the new triangle A' B' C' and give the coordinates of its corners.
2) Two similar cones have volumes of 27m³ and 64m³. If the surface area of the smaller one is 36m², find the surface area of the other one.

The Four Transformations

T ranslation — ONE Detail	
E nlargement — TWO Details	
R otation — THREE Details	
R eflection — ONE Detail	
Y	

1) Use the name TERRY to remember the 4 types.
2) You must always specify all the details for each type.
3) It'll help if you remember which properties remain unchanged in each transformation, too.

1) TRANSLATION

You must Specify this ONE detail: 1) The VECTOR OF TRANSLATION (See P.72 on vector notation)

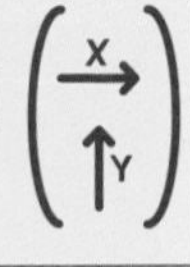

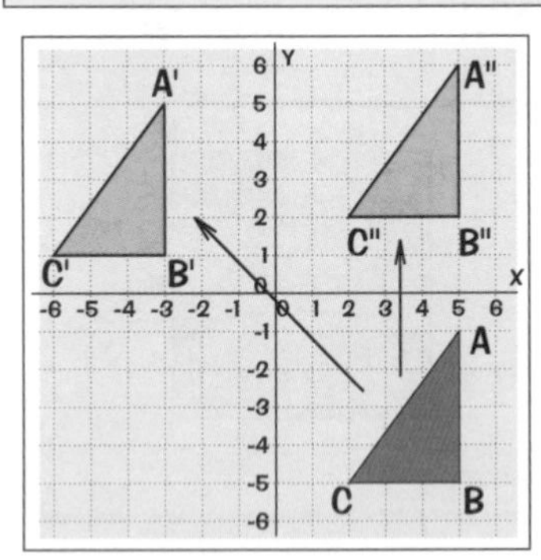

ABC to A'B'C' is a translation of $\begin{pmatrix}-8\\6\end{pmatrix}$

ABC to A"B"C" is a translation of $\begin{pmatrix}0\\7\end{pmatrix}$

All that changes in a translation is the *POSITION* of the object — *everything else* remains *unchanged*.

2) ENLARGEMENT

You must Specify these 2 details: 1) The SCALE FACTOR 2) The CENTRE of Enlargement

From A to B is an enlargement of scale factor 2, and centre (2,6)

From B to A is an enlargement of scale factor 1/2 and centre (2,6)

The *ANGLES* of the object and *RATIOS* of the lengths remain *unchanged*. The *ORIENTATION* is unchanged unless the scale factor is negative.

3) ROTATION

You must Specify these 3 details: 1) ANGLE turned 2) DIRECTION (Clockwise or..) 3) CENTRE of Rotation

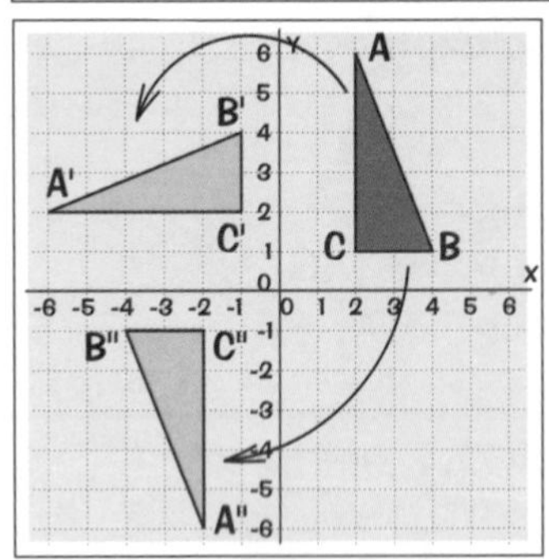

ABC to A'B'C' is a Rotation of 90°, anticlockwise, ABOUT the origin.

ABC to A"B"C" is a Rotation of half a turn (180°), clockwise, ABOUT the origin.

The only things that *change* in a rotation are the *POSITION* and the *ORIENTATION* of the object. *Everything else* remains *unchanged*.

4) REFLECTION

You must Specify this ONE detail: 1) The MIRROR LINE

A to B is a reflection in the y-axis.

A to C is a reflection in the line y = x

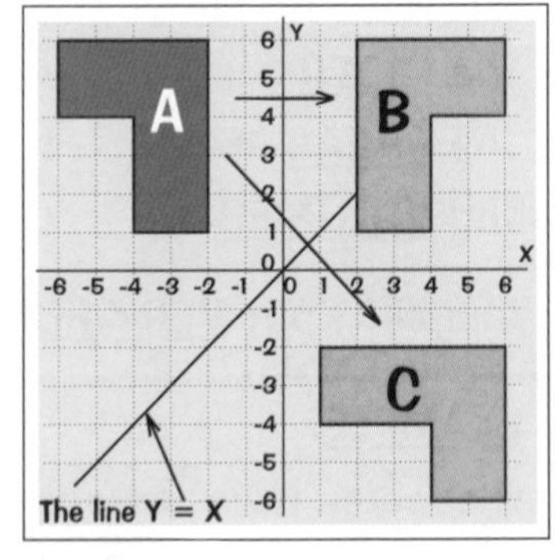

With reflection, the *POSITION* and *ORIENTATION* of the object are the *only things that change*.

The Acid Test:

LEARN the names of the Four Transformations and the details that go with each. When you think you know it, turn over and write it all down.

1) Describe *fully* these transformations: A → B, B → C, C → A, A → D.

Combinations of Transformations

In Exam questions they'll often do something horrid like stick two transformations together and then ask you what combination gets you from shape A to shape B. Be ready.

The Better You Know Them All — The Easier it is

These kinds of question aren't so bad — but ONLY if you've LEARNT the four transformations on the last page really well — if you don't know them, then you certainly won't do too well at spotting a combination of one followed by another.
That's because the method is basically "Try it and see..."

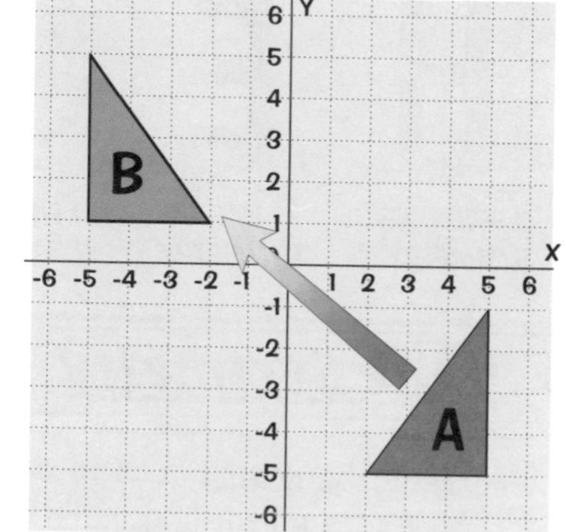

"What combination of two transformations takes you from triangle A to triangle B?"

(There's usually a few different ways of getting from one shape to the other — but remember you only need to find ONE of them.)

Method: Try an obvious transformation first, and See...

If you think about it, the answer can only be a combination of two of the four types shown on the last page, so you can immediately start to narrow it down:

1) Since the shapes are the same size we can rule out enlargements.
2) Next, try a reflection (in either the x-axis or the y-axis).
   Here we've tried a reflection in the y-axis, to give shape A':
3) You should now easily be able to see the final step
   from A' to B — it's a translation of $\begin{pmatrix}0\\6\end{pmatrix}$.

And that's it DONE — from A to B is simply a combination of:

A REFLECTION IN THE Y-AXIS followed by a TRANSLATION OF $\begin{pmatrix}0\\6\end{pmatrix}$

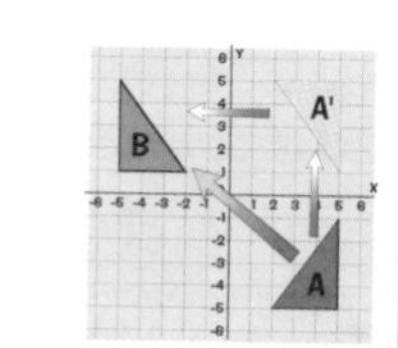

At least that's one answer anyway. If instead we decided to reflect it in the x-axis first (as shown here) then we'd get another answer (see Acid Test below) — but both are right.

"But which transformation do I try first?" I hear you cry.

Well it just depends on how it looks.
But the more transformation questions you do, the more obvious that first guess becomes.
In other words: the more you practise, the easier you'll be able to do it — surprise surprise...

The Acid Test:

LEARN the main points on this page.
Then cover it up and write them all down.

1) What pair of transformations will convert shape C into shape D?:
   What pair will convert shape D to shape C?
2) In the example above, find the other transformation needed to get to shape B after reflecting shape A in the X-axis.

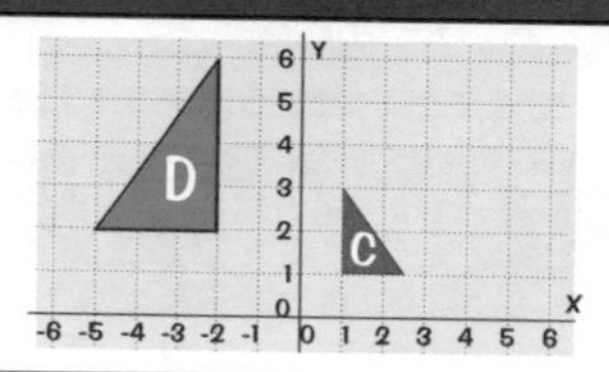

Symmetry

SYMMETRY is where a shape or picture can be put in DIFFERENT POSITIONS that LOOK EXACTLY THE SAME. There are THREE TYPES of symmetry:

1) Line Symmetry

This is where you can draw a MIRROR LINE (or more than one) across a picture and both sides will fold exactly together.

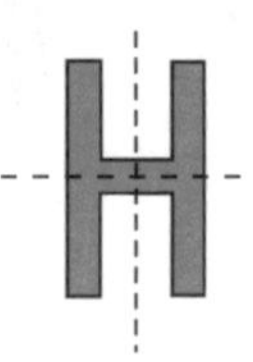

2 LINES OF SYMMETRY

1 LINE OF SYMMETRY

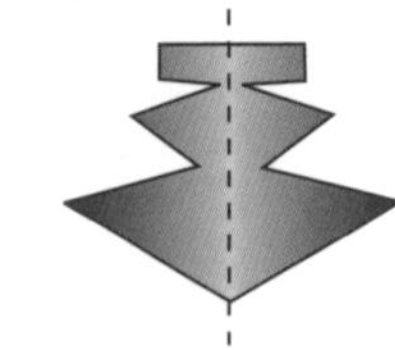

1 LINE OF SYMMETRY

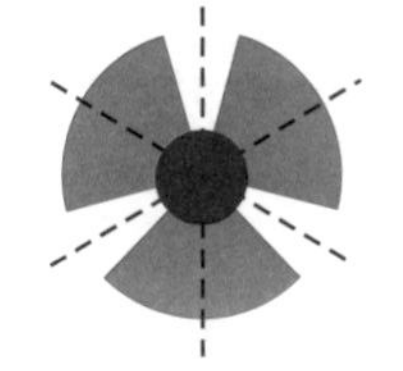

3 LINES OF SYMMETRY

NO LINES OF SYMMETRY

1 LINE OF SYMMETRY

2) Plane Symmetry

Plane Symmetry is all to do with 3-D SOLIDS. Whereas flat shapes can have a mirror line, solid 3-D objects can have planes of symmetry.

A plane mirror surface can be drawn through many regular solids, but the shape must be exactly the same on both sides of the plane (i.e. mirror images), like these are:

Planes of Symmetry

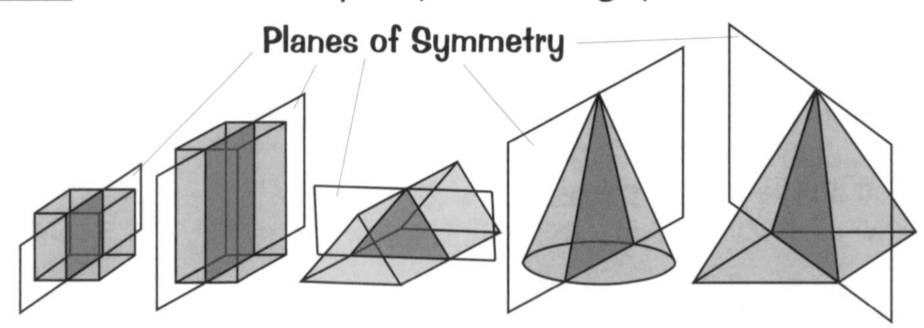

The shapes drawn here all have many more planes of symmetry but there's only one drawn in for each shape, because otherwise it would all get really messy and you wouldn't be able to see anything.

3) Rotational Symmetry

This is where you can rotate the shape into different positions that look exactly the same.
If a shape has only 1 position, you can either say "order 1 symmetry" or "no rotational symmetry"

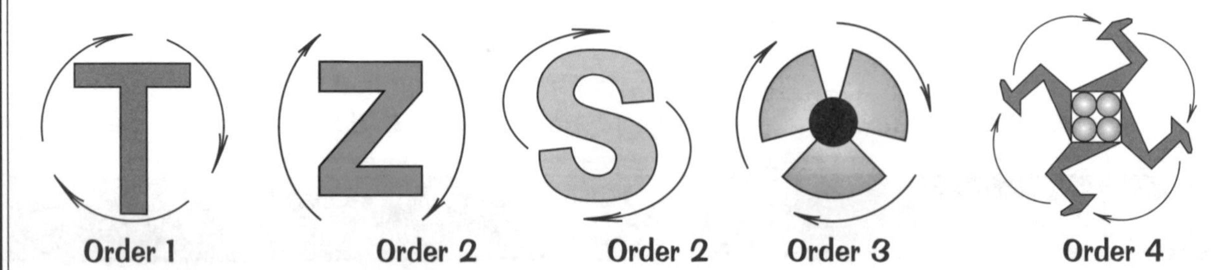

Order 1 — Order 2 — Order 2 — Order 3 — Order 4

1) Find the lines of symmetry and order of rotational symmetry for:
H N E Y M B S T

2) Work out the other planes of symmetry for the 3D solids above.

Pythagoras' Theorem and Bearings

Pythagoras' Theorem — $a^2 + b^2 = h^2$

1) PYTHAGORAS' THEOREM always goes hand in hand with SIN, COS and TAN because they're both involved with RIGHT-ANGLED TRIANGLES.
2) The big difference is that Pythagoras does not involve any angles — it just uses two sides to find the third side. (SIN, COS and TAN always involve ANGLES)
3) The BASIC FORMULA for Pythagoras is $a^2 + b^2 = h^2$
4) PLUG THE NUMBERS IN and work it out.
5) But get the numbers in the RIGHT PLACE. The 2 shorter sides (squared) add to equal the longest side (squared).
6) Always CHECK that your answer is SENSIBLE.

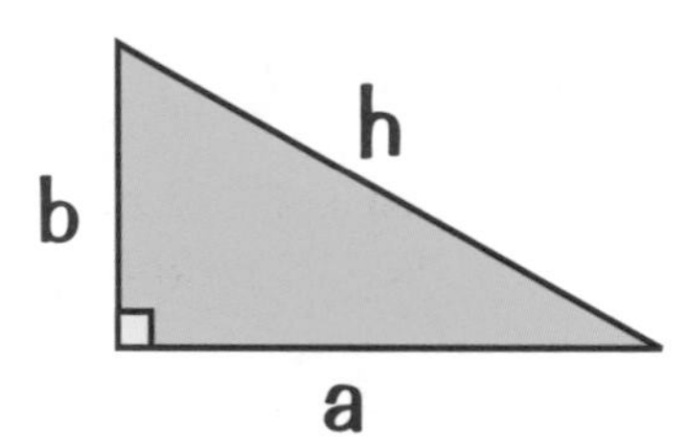

Example

Find the missing side in the triangle shown."

5m
3m

ANSWER:

$a^2 + b^2 = h^2$

$\therefore\ 3^2 + x^2 = 5^2$

$\therefore\ 9 + x^2 = 25$

$\therefore\ x^2 = 25 - 9 = 16$

$\therefore\ x = \sqrt{16} = \underline{4m}$

(Is it sensible? — Yes, it's shorter than 5m, but not too much shorter)

Bearings

To find or plot a bearing you must remember the three key words:

1) "FROM"

Find the word "FROM" in the question, and put your pencil on the diagram at the point you are going "from".

2) NORTHLINE

At the point you are going FROM, draw in a NORTHLINE.

3) CLOCKWISE

Now draw in the angle CLOCKWISE from the northline to the line joining the two points. This angle is the required bearing.

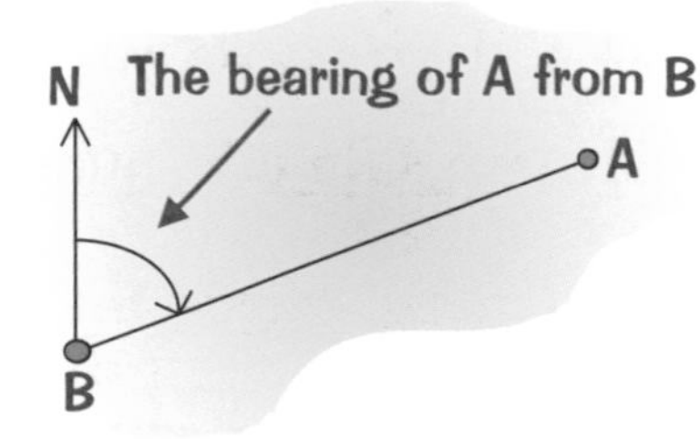

Example

"Find the bearing of Q from P":

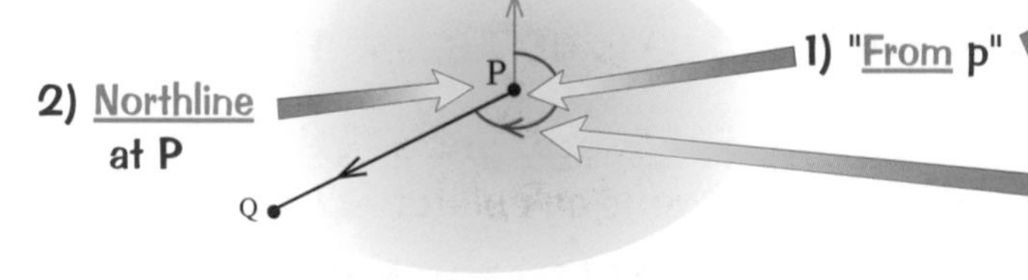

1) "From p"

3) Clockwise, from the N-line. This angle is the bearing of Q from P and is 245°.

N.B. All bearings should be given as 3 figures, e.g. 176°, 034° (not 34°), 005° (not 5°), 018° etc.

The Acid Test:

LEARN the 6 facts about Pythagoras and the 3 key words for bearings. Then turn over and write them down.

1) Find the length of BC.
2) Find the bearing of T from H, by measuring from the diagram with a protractor.
3) CALCULATE the back bearing (of H from T).

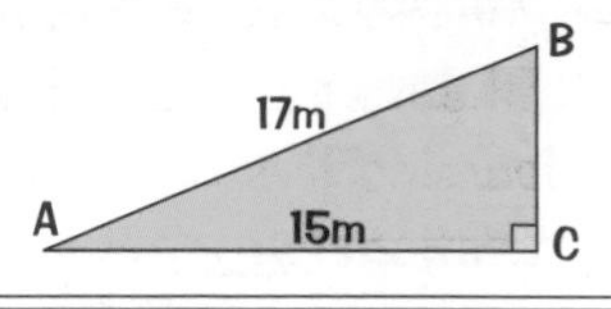

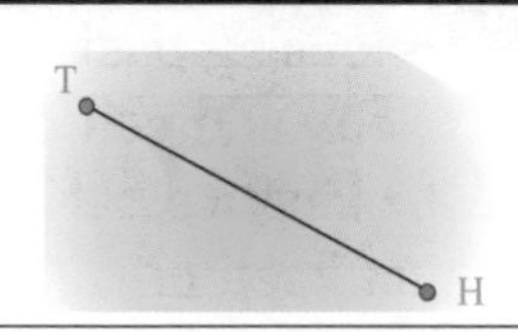

Trigonometry — Sin, Cos, Tan

There are several methods for doing Trig and they're all pretty much the same. However, the method shown below has a number of advantages, mainly because the formula triangles mean the same method is used every time, (no matter which side or angle is being asked for). This makes the whole topic a lot simpler, and you'll find that once you've learned this method, the answers automatically come out right every time. It's just a joy.

Method

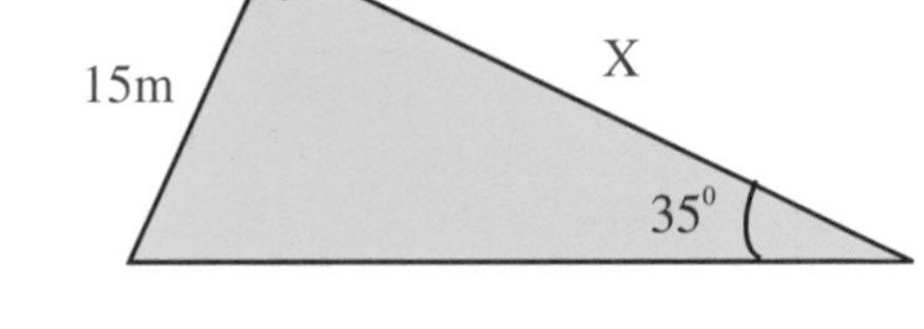

1) Label the three sides O, A and H (Opposite, Adjacent and Hypotenuse).
2) Write down from memory "SOH CAH TOA" (Sounds like a Chinese word, "Sockatoa!")
3) Decide which two sides are involved: O,H A,H or O,A and select SOH, CAH or TOA accordingly
4) Turn the one you choose into a FORMULA TRIANGLE:

SOH CAH TOA

O / Sθ X H

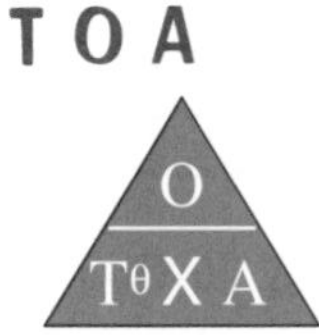

5) Cover up the thing you want to find (with your finger), and write down whatever is left showing.

6) Translate into numbers and work it out.
7) Finally, check that your answer is sensible.

Some Nitty Gritty Details

1) The Hypotenuse is the LONGEST SIDE. The Opposite is the side OPPOSITE the angle being used (θ), and the Adjacent is the other side NEXT TO the angle being used.
2) In the formula triangles, Sθ represents SIN θ, Cθ is COS θ, and Tθ is TAN θ.
3) Remember, TO FIND THE ANGLE — USE INVERSE. i.e. press INV or SHIFT or 2nd, followed by SIN, COS or TAN (and make sure your calculator is in DEG mode).
4) You can only use SIN, COS and TAN on RIGHT-ANGLED TRIANGLES — you may have to add lines to the diagram to create one, especially with isosceles triangles.

The Acid Test:

LEARN the 7 Steps of the Method and the Four Nitty Gritty Details. Then turn over and write them down.

Practising past paper questions is very important, but the whole point of doing so is to check and consolidate the methods you have already learnt. Don't make the mistake of thinking it's pointless learning these 7 steps. If you don't know them all thoroughly, you'll just keep on getting questions wrong.

Trigonometry — Sin, Cos, Tan

Example 1) "Find x in the triangle shown."

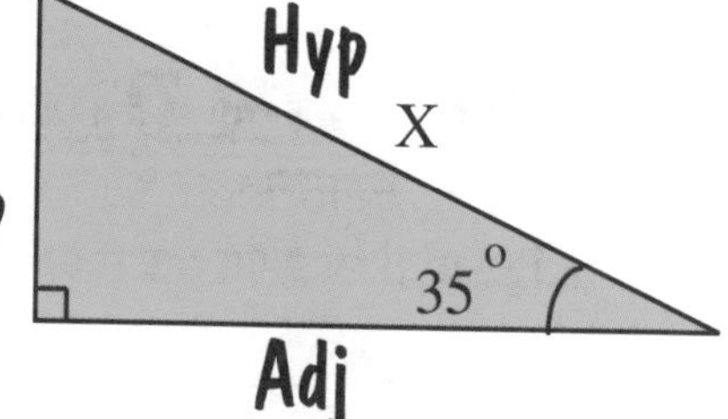

1) Label O,A,H
2) Write down "SOH CAH TOA"
3) Two sides involved: O,H
4) So use 

5) We want to find H so cover it up to leave: $h = \frac{O}{s\theta}$
6) Translate: $x = \frac{15}{\sin 35}$

   Press 15 ÷ SIN 35 = 26.151702 **So ans = 26.2m**
7) Check it's sensible: yes it's about twice as big as 15, as the diagram suggests.

Example 2) "Find the angle θ in this triangle."

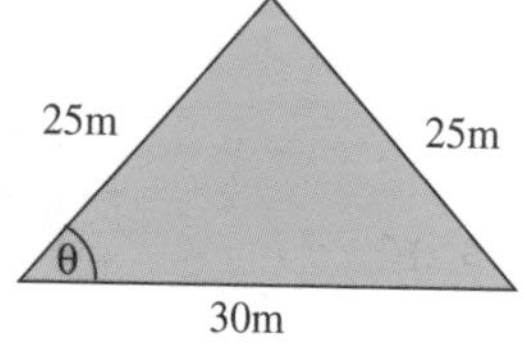

Note the usual way of dealing with an ISOSCELES TRIANGLE: split it down the middle to get a RIGHT ANGLE:

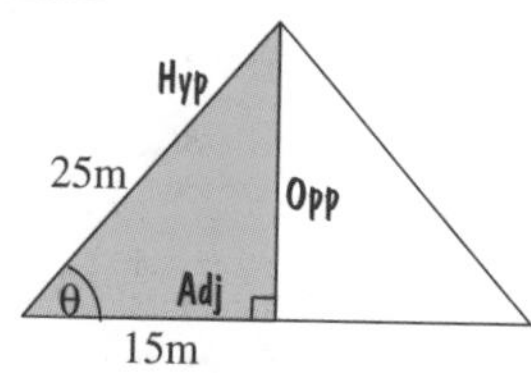

1) Label O, A, H
2) Write down "SOH CAH TOA"
3) Two sides involved: A,H
4) So use 

5) We want to find θ so cover up Cθ to leave: $c\theta = \frac{A}{H}$
6) Translate: $\cos\theta = \frac{15}{25} = 0.6$

   NOW USE INVERSE: θ = INV COS (0.6)

   Press 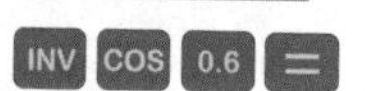 53.130102 **So ans. = 53.1°**

7) Finally, is it sensible? — Yes, the angle looks like about 50°.

Angles of Elevation And Depression

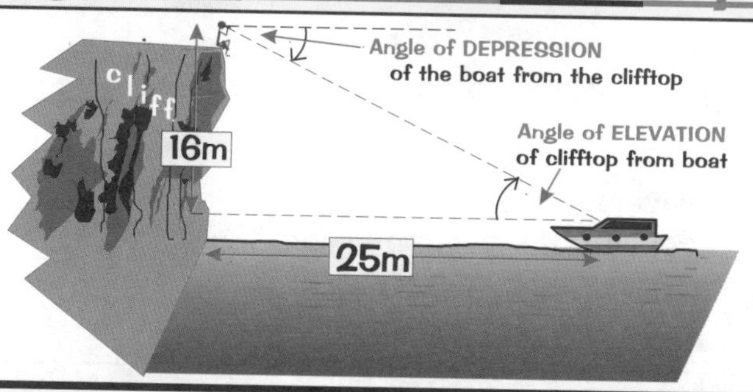

1) The Angle of Depression is the angle downwards from the horizontal.
2) The Angle of Elevation is the angle upwards from the horizontal.
3) The Angles of Elevation and Depression are EQUAL.

The Acid Test:

Practise these three questions until you can apply the method fluently and without having to refer to it at all.

1) Find x
2) Find θ

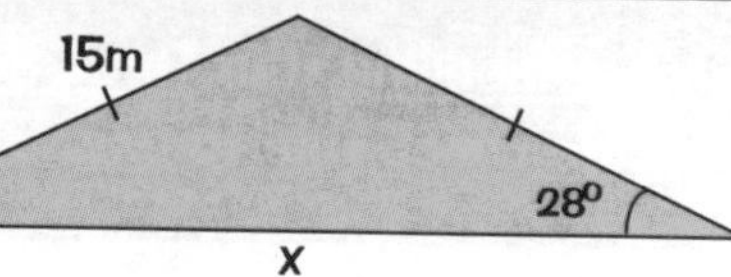

15m, 6m, θ

3) Calculate the angles of elevation and depression in the boat drawing above.

The Sine and Cosine Rules

Normal trigonometry using SOH CAH TOA etc. can only be applied to right-angled triangles. The Sine and Cosine Rules on the other hand allow you to tackle any triangle at all with ease.

Labelling The Triangle

This is very important. You must label the sides and angles properly so that the letters for the sides and angles correspond with each other:

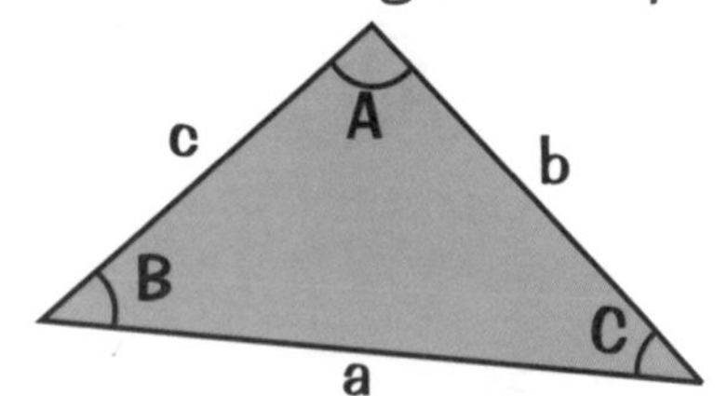

Remember, side "a" is opposite angle A etc.

It doesn't matter which sides you decide to call a, b, and c, just as long as the angles are then labelled properly.

Three Formulas to Learn:

These first two formulas let you work out sides and angles:

The Sine Rule

$$\frac{a}{\text{SIN A}} = \frac{b}{\text{SIN B}} = \frac{c}{\text{SIN C}}$$

You don't use the whole thing with both "=" signs of course, so it's not half as bad as it looks — you just choose the two bits that you want: e.g. $\frac{b}{\text{SIN B}} = \frac{c}{\text{SIN C}}$ or $\frac{a}{\text{SIN A}} = \frac{b}{\text{SIN B}}$

The Cosine Rule

$$a^2 = b^2 + c^2 - 2bc\,\text{COS A}$$

or $$\text{COS A} = \frac{b^2 + c^2 - a^2}{2bc}$$

Area of the Triangle

Of course, you already know the simple formula when you have the base and vertical height:

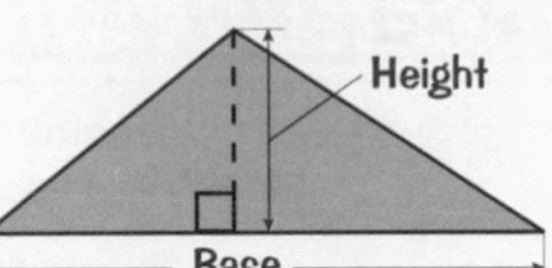

Area = ½ base × height

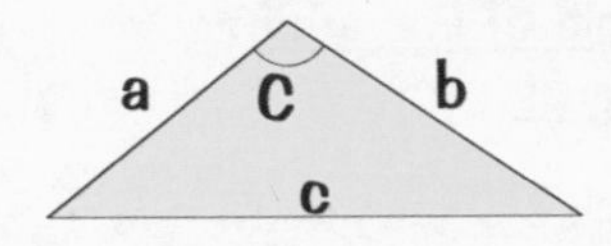

Well, here's a fancier formula that you can use when you know two sides and the angle between them:

Area of triangle = ½ abSINC

You need to LEARN all of these formulas off by heart and practise using them. If you don't, you won't be able to use them in the Exam, even if they give them to you.

The Acid Test:

LEARN the proper labelling, the Three Formulas, and how to decide which rule to use.

Now turn over and write down everything on this page.

The Sine and Cosine Rules

The Four Examples

Amazingly enough there are basically only FOUR questions where the SINE and COSINE rules would be applied. Learn the exact details of these four basic examples:

1) TWO ANGLES given plus ANY SIDE:

— SINE RULE NEEDED

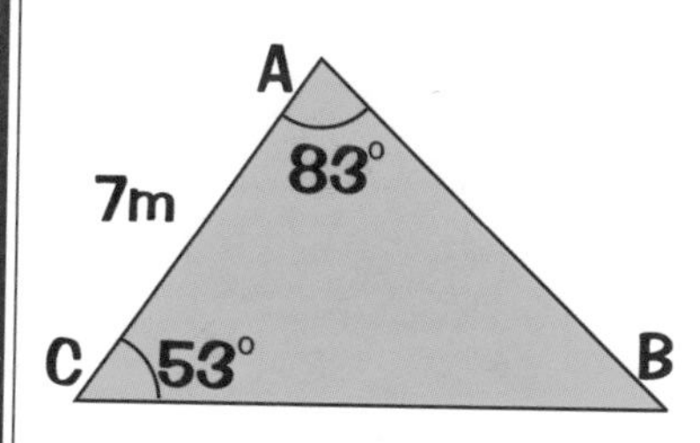

1) Don't forget the obvious: $B = 180 - 83 - 53 = \underline{44^\circ}$

2) Then use $\frac{b}{\text{SIN}\,B} = \frac{c}{\text{SIN}\,C} \Rightarrow \frac{7}{\text{SIN}\,44} = \frac{c}{\text{SIN}\,53}$

3) Which gives $\Rightarrow c = \frac{7 \times \text{SIN}\,53}{\text{SIN}\,44} = \underline{8.05\text{m}}$

The rest is easy using the SINE RULE

2) TWO SIDES given plus an ANGLE NOT ENCLOSED by them

— SINE RULE NEEDED

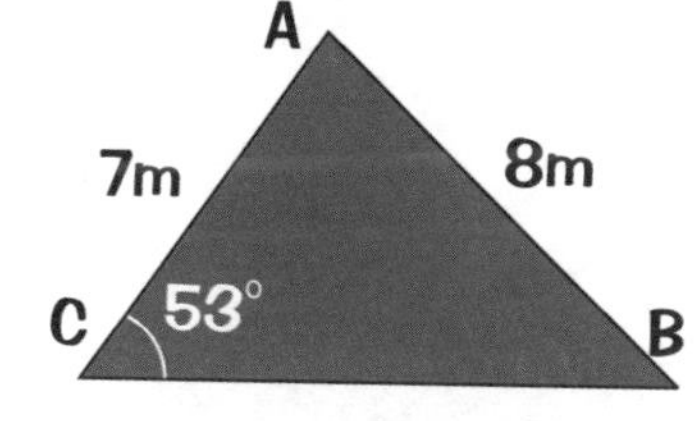

1) Use: $\frac{b}{\text{SIN}\,B} = \frac{c}{\text{SIN}\,C} \Rightarrow \frac{7}{\text{SIN}\,B} = \frac{8}{\text{SIN}\,53}$

2) $\Rightarrow \text{SIN}\,B = \frac{7 \times \text{SIN}\,53}{8} = 0.6988 \Rightarrow B = \text{SIN}^{-1}(0.6988) = 44.3^\circ$

The rest is easy using the SINE RULE

3) TWO SIDES given plus THE ANGLE ENCLOSED by them

— COSINE RULE NEEDED

1) Use: $a^2 = b^2 + c^2 - 2bc\,\text{COS}\,A$

$= 7^2 + 8^2 - 2 \times 7 \times 8 \times \text{COS}\,83$

$= 99.3506 \Rightarrow a = \sqrt{99.3506} = \underline{9.97\text{m}}$

The rest is easy using the SINE RULE

4) ALL THREE SIDES given but NO ANGLES

— COSINE RULE NEEDED

1) Use: $\text{COS}\,A = \frac{b^2 + c^2 - a^2}{2bc}$

$= \frac{49 + 64 - 100}{2 \times 7 \times 8} = \frac{13}{112} = 0.11607$

2) Hence $A = \text{COS}^{-1}(0.11607) = \underline{83.3^\circ}$

The rest is easy using the SINE RULE

The Acid Test:

LEARN the FOUR BASIC TYPES as above. Then cover the page and do these:

1) Write down a new version of each of the 4 examples above and then use the SINE and COSINE RULES to find ALL of the sides and angles for each one.
2) A triangle has two sides of 12m and 17m with an angle of 70° between them. Find all the other sides and angles in the triangle. (A sketch is essential, of course).

Vectors

Three monstrously important things you need to know about vectors:

1) The Four Notations

The vector shown here can be referred to as:

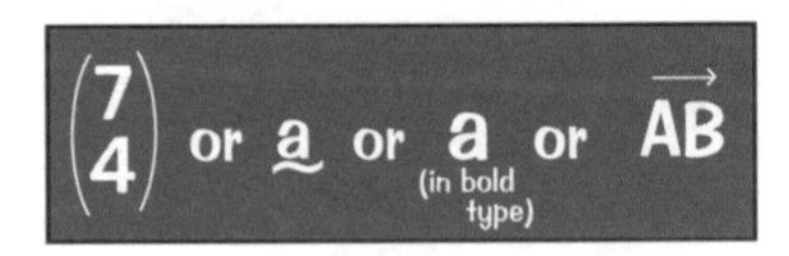

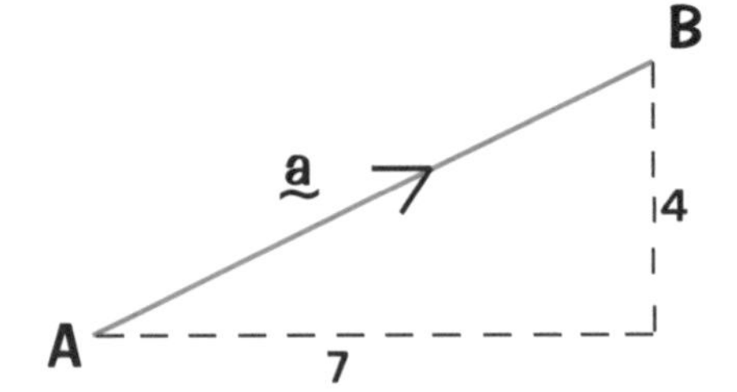

It's pretty obvious what these mean. Just make sure you know which is which in the column vector (x→ and y↑) and what a negative value means in a column vector.

2) Adding And Subtracting Vectors

Vectors must always be added end to end, so that the arrows all point with each other, not against each other.

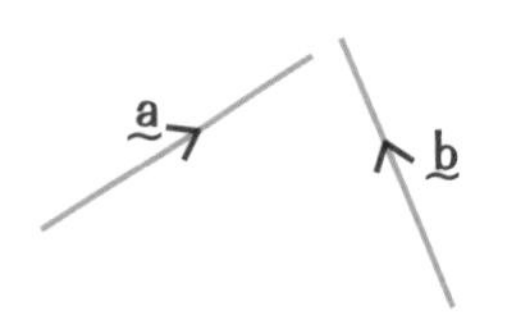

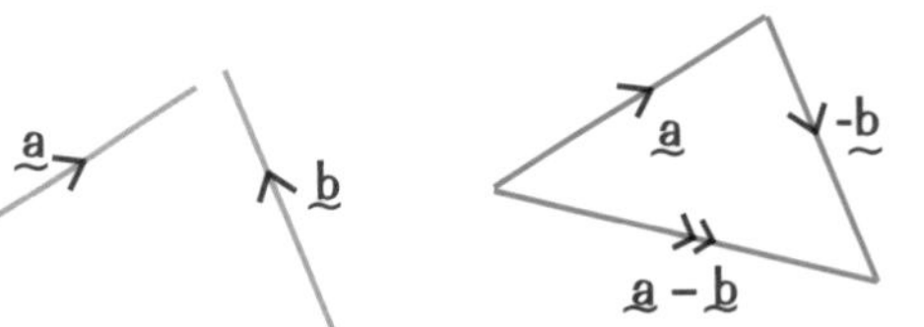

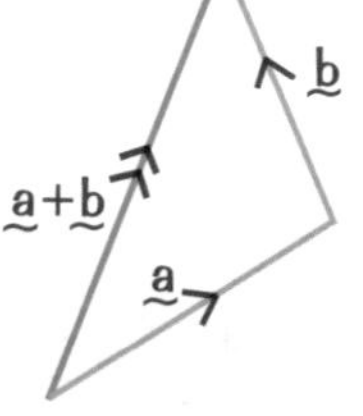

Adding and subtracting COLUMN VECTORS is really easy:

E.g. if $a = \begin{pmatrix}5\\3\end{pmatrix}$ and $b = \begin{pmatrix}-2\\4\end{pmatrix}$ then $2a - b = 2\begin{pmatrix}5\\3\end{pmatrix} - \begin{pmatrix}-2\\4\end{pmatrix} = \begin{pmatrix}12\\2\end{pmatrix}$

3) A Typical Exam Question

This is a common type of question and it illustrates a very important vector technique:

To obtain the unknown vector just 'get there' by any route made up of known vectors

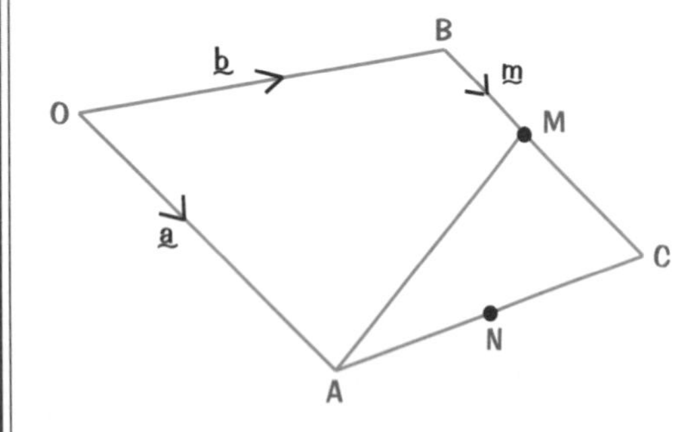

Applying this rule we can easily obtain the following vectors in term of $\underset{\sim}{a}$, $\underset{\sim}{b}$ and $\underset{\sim}{m}$ (given that M and N are mid points):

1) $\overrightarrow{AM} = -\underset{\sim}{a} + \underset{\sim}{b} + \underset{\sim}{m}$ (i.e. get there via O and B)

2) $\overrightarrow{OC} = \underset{\sim}{b} + 2\underset{\sim}{m}$ (i.e. get there via B and M)

3) $\overrightarrow{AC} = -\underset{\sim}{a} + \underset{\sim}{b} + 2\underset{\sim}{m}$ (A to C via O, B and M)

The Acid Test:

LEARN the important details on this page, then turn over and write them down.

1) For the diagram above, express the following in terms of $\underset{\sim}{a}$, $\underset{\sim}{b}$ and $\underset{\sim}{m}$:

a) $\overrightarrow{MO}$ b) $\overrightarrow{AN}$ c) $\overrightarrow{BN}$ d) $\overrightarrow{NM}$

"Real Life" Vector Questions

This page covers the trickier types of vector question you could get. Make sure you learn all the little tricks.

1) The Old "Swimming Across the River" Question

This is a really easy question: You just ADD the two velocity vectors END TO END and draw the RESULTANT VECTOR which shows both the speed and direction of the final course. Simple huh?

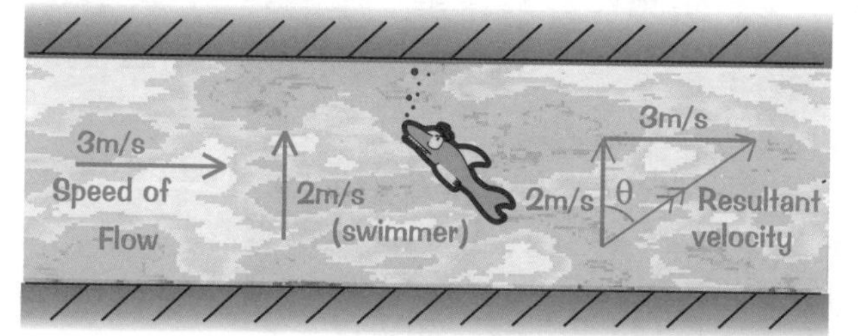

Overall Speed = $\sqrt{3^2 + 2^2} = \sqrt{13} = 3.6\text{m/s}$

Direction: $\tan\theta = 3 \div 2$

$\theta = \tan^{-1}(1.5) = 56.3^\circ$

As usual with vectors, you'll need to use Pythagoras and Trig to find the length and angle but that's no big deal is it? Just make sure you LEARN the two methods in this question.
The example shown above is absolutely dog-standard stuff and you should definitely see it that way, rather than as one random question of which there may be hundreds — there aren't!

2) The Old "Swimming Slightly Upstream" Question

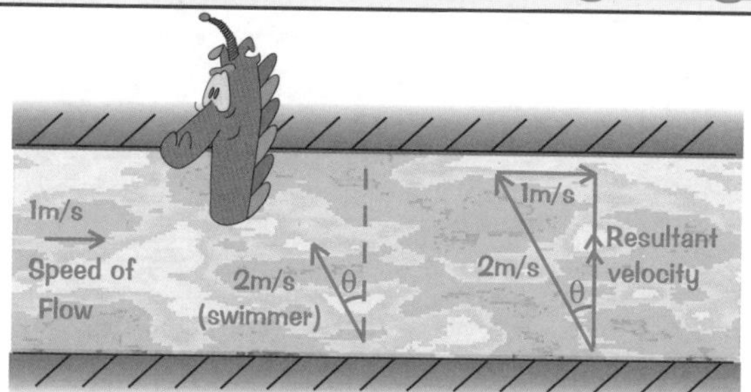

1) $\sin\theta = \text{OPP/HYP}$

$= 1/2$

so $\theta = \sin^{-1}(0.5) = 30^\circ$

2) Speed $= \sqrt{2^2 - 1^2} = \sqrt{3} =$ 1.73 m/s

The general idea here is to end up going directly across the river, and once again the old faithful method of drawing a vector triangle makes light work of the whole thing — 2 vectors joined end to end to give the resultant velocity. However, in this case the resultant is drawn in first (straight across), so that the angle θ has to be worked out to fit as shown above.

3) The Old "Queen Mary's Tugboats" Question

The problem here is to find the overall force from the two tugs.

LINER

Tug A 10,000N

40°

20°

Tug B 15,000N

This is tackled by adding the vectors end to end to produce a triangle like this:

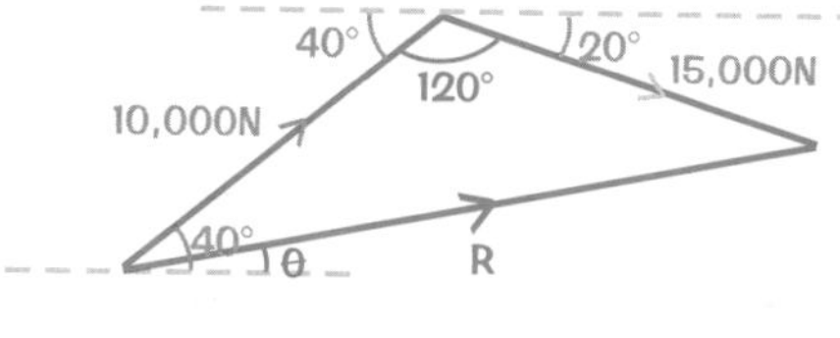

The angle 120° inside the triangle is found using the rules for alternate angles and angles on a straight line — see P36-37.

You then need to use the SINE & COSINE RULES to find R and θ (the size and direction of the resultant force).

The Acid Test: LEARN the 3 EXAMPLES on this page, then turn over and write them out, but with different numbers.

1) Work out the overall force on the Queen Mary in example 3 (by finding R and θ).

3D Pythagoras and Trigonometry

3D questions on Pythagoras and trig might seem a bit mind-boggling at first — but you're really just using those same old rules.

Angle Between Line and Plane — Use a Diagram

Learn The 3-Step Method

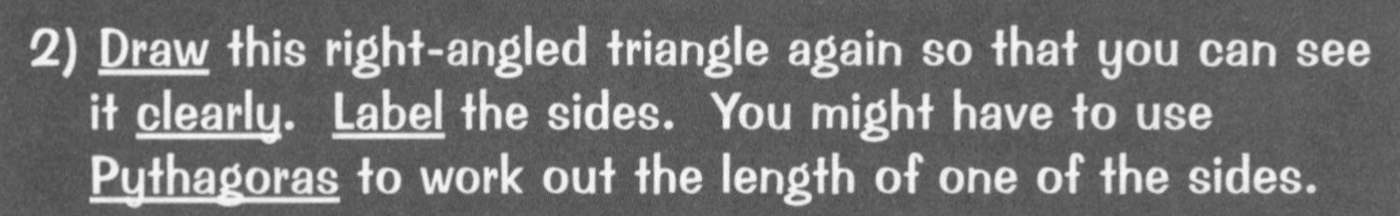

1) Make a RIGHT-ANGLED triangle using the line, a line in the plane and a line between the two.
2) Draw this right-angled triangle again so that you can see it clearly. Label the sides. You might have to use Pythagoras to work out the length of one of the sides.
3) Use trigonometry to calculate the angle.

Example: "ABCDE is a square-based pyramid. It is 12 cm high and the square base has sides of length 7 cm. Find the angle the edge AE makes with the base."

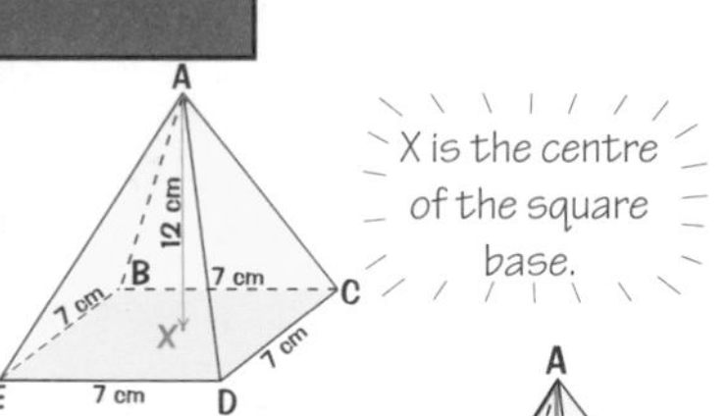

1) First draw a right-angled triangle using the edge AE, the base and a line between the two (in this case the central height). Call the angle you're trying to find θ.

2) Now draw this triangle clearly and label it.

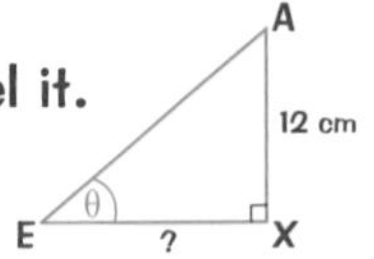

To find θ, you need to know the length of side EX.

So, using Pythagoras — $EX^2 = 3.5^2 + 3.5^2 = 24.5 \Rightarrow EX = \sqrt{24.5}$ cm

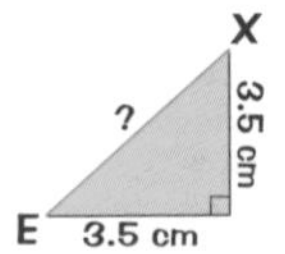

You know the lengths of the opposite and adjacent sides, so use tan.

3) Now use trigonometry to find the angle θ:

$$\tan\theta = \frac{12}{\sqrt{24.5}} = 2.4... \quad \theta = 67.6° \text{ (1 d.p.)}$$

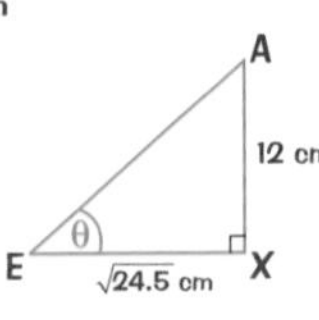

Use Right-Angled Triangles To Find Lengths too

Example:

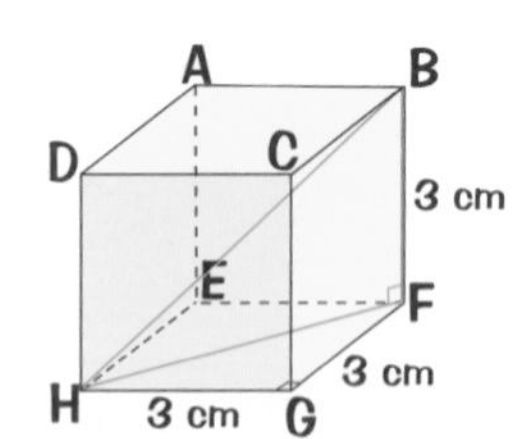

1) First use Pythagoras to find the length FH.

$FH^2 = 3^2 + 3^2 = 18 \Rightarrow FH = \sqrt{18}$ cm

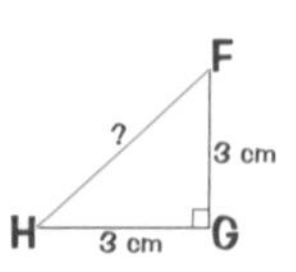

2) Now use Pythagoras again to find the length BH.

$BH^2 = 3^2 + (\sqrt{18})^2 = 27 \Rightarrow BH = \sqrt{27}$ cm $=$ 5.2 cm (1 decimal place)

The Acid Test:

LEARN THE 3-STEP METHOD — then try the questions below.

1) Calculate the angle that the line AG makes with the base of this cuboid.
2) Calculate the length of AG.

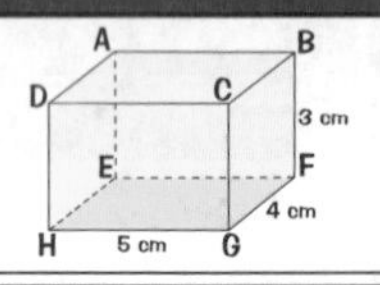

Straight Line Graphs: Gradient

Strict Method for Finding Gradient

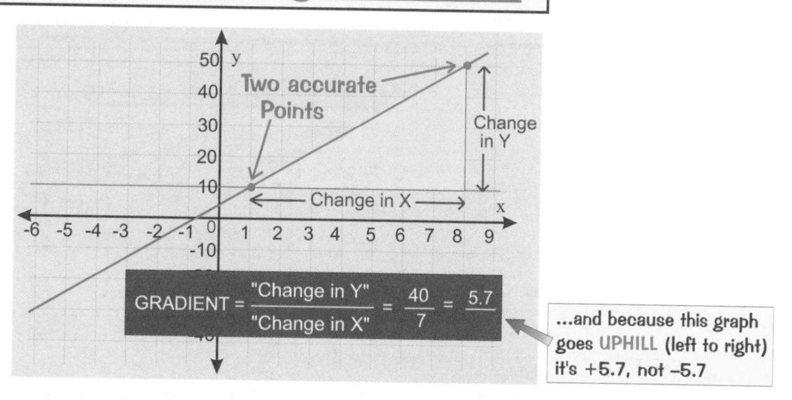

...and because this graph goes UPHILL (left to right) it's +5.7, not –5.7

1) Find TWO ACCURATE POINTS and COMPLETE THE TRIANGLE

Both points in the upper right quadrant if possible (to keep all the numbers positive).

2) Find the CHANGE IN Y and the CHANGE IN X

Make sure you subtract the x coords. the same way round as you do the y coords.
E.g. y coord. of pt A – y coord. of pt B and x coord of pt A – x coord of pt B

3) LEARN this formula, and use it:

$$\text{GRADIENT} = \frac{\text{CHANGE IN Y}}{\text{CHANGE IN X}}$$

4) Check the SIGN'S right.

If it slopes UPHILL left → right (↗) then it's positive
If it slopes DOWNHILL left → right (↘) then it's negative

If you subtracted the coordinates the right way round, the sign should be correct. If it's not, go back and check what you've done.

Parallel and Perpendicular Lines

1) The equation of a straight line is $y = mx + c$, where m is the gradient and c is the y-intercept.

2) Parallel lines have the same value of m, i.e. the same gradient. So the lines: $y = 2x + 3$, $y = 2x$ and $y = 2x - 4$ are all parallel.

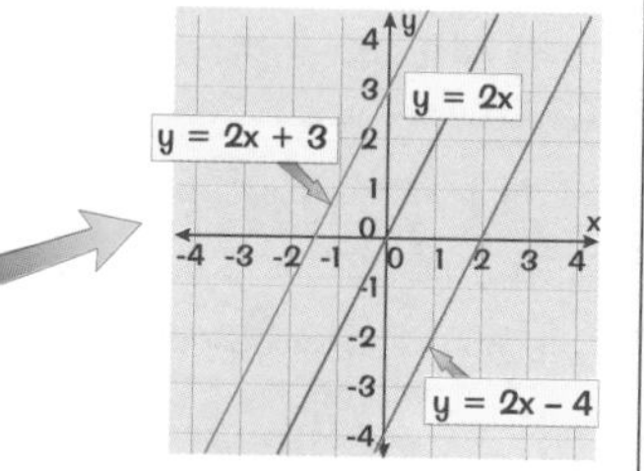

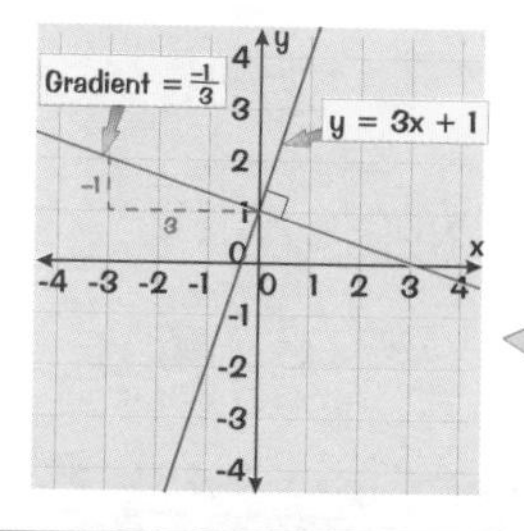

3) The gradients of two perpendicular lines multiply to give –1.

If the gradient of the first line is m, the gradient of the other line will be $\frac{-1}{m}$, because $m \times \frac{-1}{m} = -1$.

The Acid Test:

LEARN the FOUR STEPS for finding a gradient then turn over and WRITE THEM DOWN from memory.

1) Plot these 3 points on a graph: (0,3) (2,0) (5,-4.5) and then join them up with a straight line. Now carefully apply the FOUR STEPS to find the gradient of the line.

Straight Line Graphs: "Y = mx + c"

Using "$y = mx + c$" is perhaps the "proper" way of dealing with straight line equations, and it's a nice trick if you can do it. The first thing you have to do though is rearrange the equation into the standard format like this:

Straight line:		Rearranged into "$y = mx + c$"	
$y = 2 + 3x$	→	$y = 3x + 2$	($m=3$, $c=2$)
$2y - 4x = 7$	→	$y = 2x + 3\frac{1}{2}$	($m=2$, $c=3\frac{1}{2}$)
$x - y = 0$	→	$y = x + 0$	($m=1$, $c=0$)
$4x - 3 = 5y$	→	$y = 0.8x - 0.6$	($m=0.8$, $c=-0.6$)
$3y + 3x = 12$	→	$y = -x + 4$	($m=-1$, $c=4$)

REMEMBER: "m" equals the gradient of the line.
"c" is the "y-intercept" (where the graph hits the y-axis).

BUT WATCH OUT: people mix up "m" and "c" when they get something like $y = 5 + 2x$. Remember, "m" is the number in front of the "x" and "c" is the number on its own.

1) Sketching a Straight Line using y = mx + c

1) Get the equation into the form "$y = mx + c$".
2) Put a dot on the y-axis at the value of c.
3) Then go along one unit and up or down by the value of m and make another dot.
4) Repeat the same "step" in both directions.
5) Finally check that the gradient looks right.

Y = 2X +1
START HERE
m=+2 so it's UP 2 units
ALWAYS along ONE Unit →

The graph shows the process for the equation "$y = 2x + 1$":
1) "c" = 1, so put a first dot at $y = 1$ on the y-axis.
2) Go along 1 unit → and then up by 2 because "m" = +2.
3) Repeat the same step, 1→ 2↑ in both directions.
4) CHECK: a gradient of +2 should be quite steep and uphill left to right which it is, so it looks OK.

2) Finding the Equation Of a Straight Line Graph

This is the reverse process and is EASIER

1) From the axes, identify the two variables (e.g. "x and y" or "h and t").
2) Find the values of "m" (gradient) and "c" (y-intercept) from the graph.
3) Using these values from the graph, write down the equation with the standard format "$y = mx + c$".

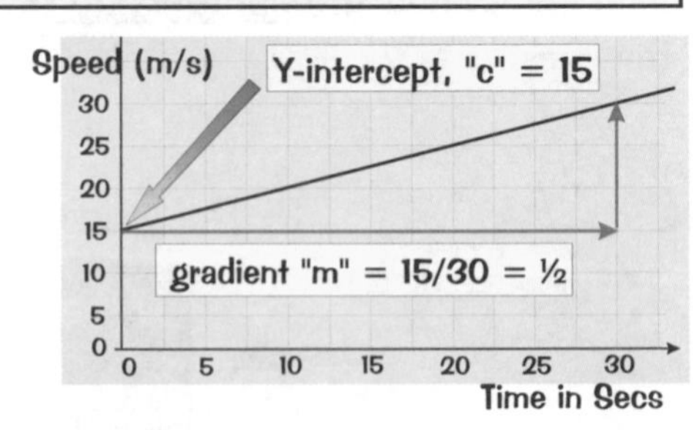

For the example above: "$S = \frac{1}{2}t + 15$"

The Acid Test:

LEARN what straight line equations look like and the 8 RULES for drawing the lines and finding the equations.

1) Sketch these graphs: a) $y = 2 + x$ b) $y = x + 6$ c) $4x - 2y = 0$ d) $y = 1 - \frac{1}{2}x$ e) $x = 2y + 4$ f) $2x - 6y - 8 = 0$ g) $0.4x - 0.2y = 0.5$ h) $y = 3 - x + 2$

Pythagoras, Lines and Line Segments

Use Pythagoras to find the Distance Between Points

You need to know how to find the straight-line distance between two points — the trick is to remember Pythagoras...

EXAMPLE: "Point P has coordinates (8, 3) and point Q has coordinates (-4, 8). Find the length of the line PQ."

If you get a question like this, follow these rules and it'll all become breathtakingly simple:

1) Draw a sketch to show the right-angled triangle.
2) Find the lengths of the sides of the triangle.
3) Use Pythagoras to find the length of the diagonal. (That's your answer.)

SOLUTION: ①

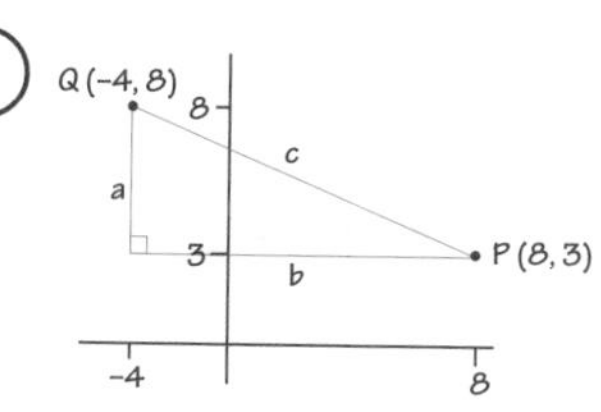

② Length of side a = 8 – 3 = 5
Length of side b = 8 – -4 = 12

③ Use Pythagoras to find side c:
$c^2 = a^2 + b^2 = 5^2 + 12^2 = 25 + 144 = 169$
So: $c = \sqrt{169} = 13$

Lines and Line Segments...

1) The example above asked you to find the length of the line PQ. To be really precise, the line PQ isn't actually a line — it's a line segment. Confused, read on...

2) A line is straight and continues to infinity (it goes on forever) in both directions. A line segment is just part of a line — it has 2 end points.

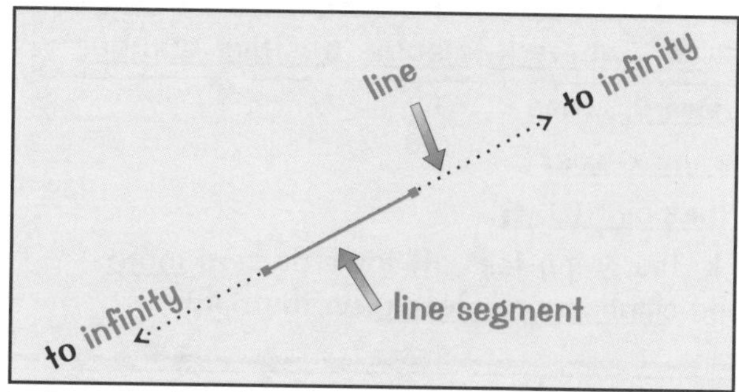

3) So the length PQ is just a chunk of the line running through P and Q. Don't worry too much about lines and line segments. The syllabus says you need to know the difference — and now you do. Just be aware that things which are actually line segments will often be referred to as lines.

The Acid Test:

LEARN the details on this page, then turn over and write them down.

1) Point A has coordinates (10, 15) and point B has coordinates (6, 12). Find the length of the line AB.
2) What is the difference between a line and a line segment?

Some Harder Graphs to Learn

There are five graphs that you should know the basic shape of just from looking at their equations — it really isn't as difficult as it sounds.

1) x^2 BUCKET SHAPES: $y = ax^2 + bx + c$ (where b and/or c can be zero)

Notice that all these graphs have the same symmetrical bucket shape and that if the x^2 bit has a "–" in front of it then the bucket is upside down.

$y = x^2$

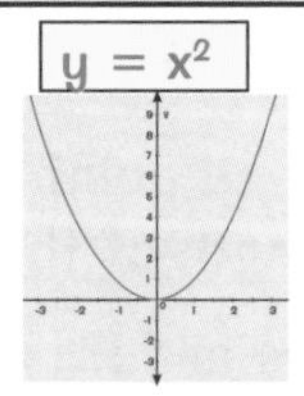

$y = 3x^2 - 6x - 3$

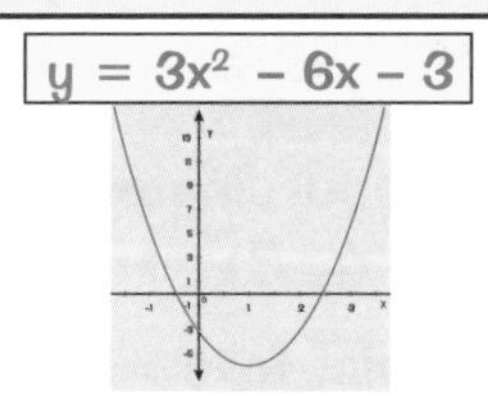

$y = -2x^2 - 4x + 3$

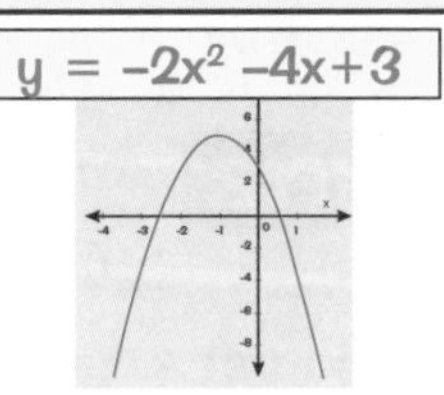

2) x^3 GRAPHS: $y = ax^3 + bx^2 + cx + d$ (Note that b, c and/or d can be zero)

(Note that x^3 must be the highest power and there must be no other bits like $\frac{1}{x}$ etc.)

All x^3 graphs have the same basic wiggle in the middle, but it can be a flat wiggle or a more pronounced wiggle. Notice that "$-x^3$ graphs" always come down from top left whereas the $+x^3$ ones go up from bottom left.

$y = x^3$

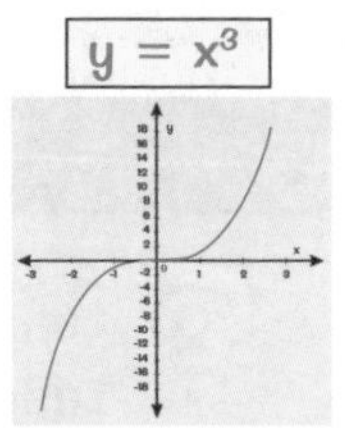

$y = x^3 + 3x^2 - 4x$

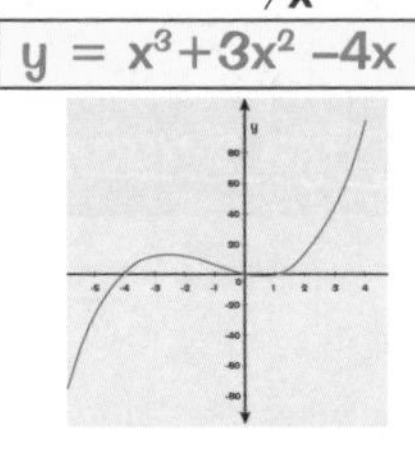

$y = -7x^3 - 7x^2 + 42x$

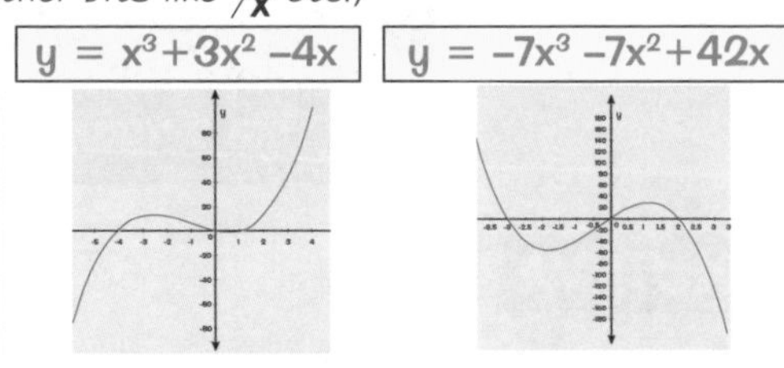

3) 1/x GRAPHS: $y = \frac{A}{x}$, or $xy = A$, where A is some number (+ or –)

$y = 4/x$ or $xy = 4$

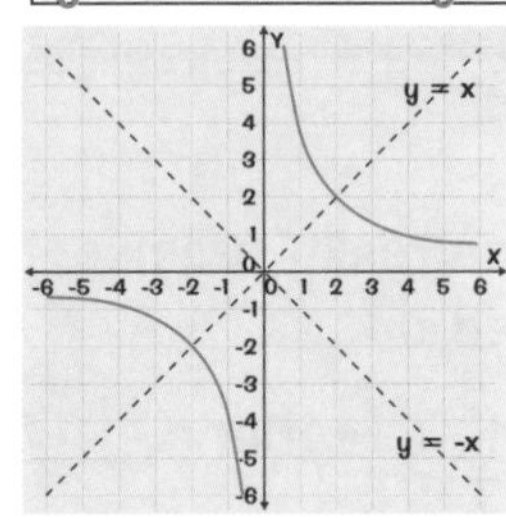

These graphs are all the same basic shape, except that the negative ones are in the opposite quadrants to the positive ones (as shown). The two halves of the graph don't touch. They're all symmetrical about the lines $y=x$ and $y=-x$. This is also the type of graph you get with inverse proportion. (See P.99)

$y = -4/x$ or $xy = -4$

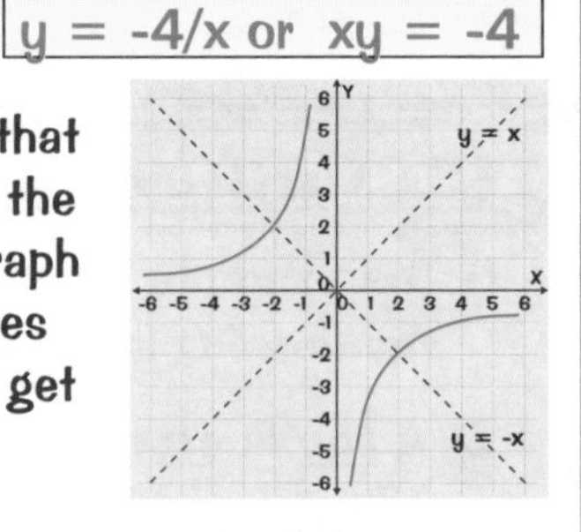

4) k^x GRAPHS: $y = k^x$, where k is some positive number

1) These graphs curve upwards when $k > 1$.
2) They're always above the x-axis.
3) They all go through the point (0, 1).
4) For bigger values of k, the graph tails off towards zero more quickly on the left and climbs more steeply on the right.

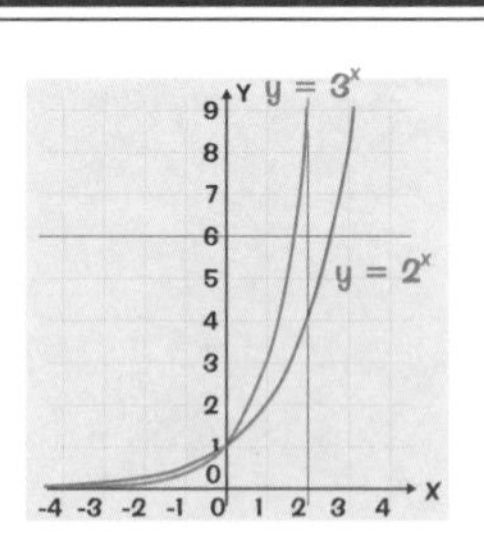

5) CIRCLES: $x^2 + y^2 = r^2$

The equation for a circle with centre (0, 0) and radius r is: $x^2 + y^2 = r^2$

$x^2 + y^2 = 25$ is a circle with centre (0, 0). $r^2 = 25$, so the radius, r, is 5.
$x^2 + y^2 = 100$ is a circle with centre (0, 0) $r^2 = 100$, so the radius, r, is 10.

$x^2 + y^2 = 25$

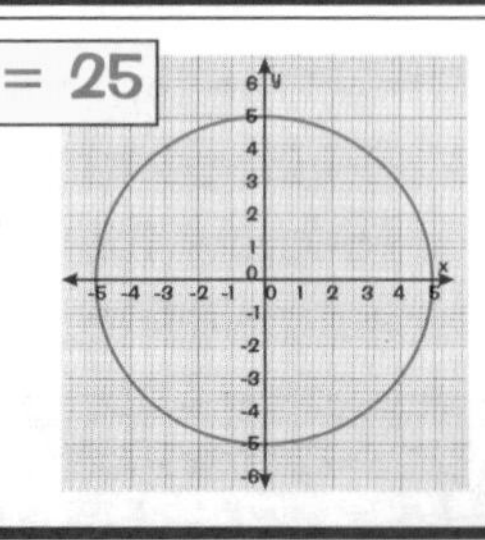

The Acid Test:

LEARN the 5 Types of Graph, both their equations and their shapes, then turn over and sketch three examples of each.

1) Describe the following graphs in words: a) $y = 3x^2 + 2$ b) $y = 4 - x^3$ c) $yx = 2$ d) $x^2 + y^2 = 36$ e) $x = -7/y$ f) $3x^2 = y - 4x^3 + 2$ g) $y = x - x^2$ h) $y = 5^x$

Quadratic Graphs

Quadratic functions are of the form $y = ax^2 + bx + c$ (where b and c can be zero) and they always have a SYMMETRICAL bucket shape — see previous page.

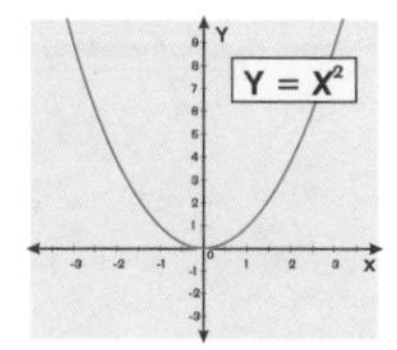

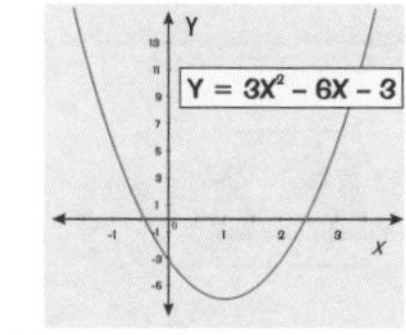

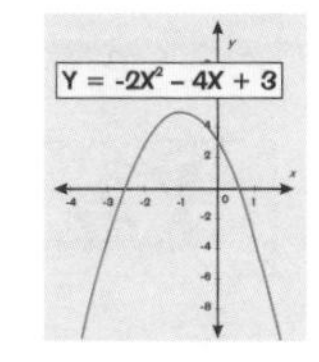

So when you plot a quadratic, remember that you're aiming for a symmetrical bucket shape — anything else is a sure sign that you've gone wrong.

Here's how to tackle questions on quadratics.

1) Fill in The Table of Values

Example: "Fill in the table of values for the equation $y = x^2 + 2x - 3$ and draw the graph."

x	-5	-4	-3	-2	-1	0	1	2	3
y		5		-3	-4	-3	0		

Work out each point very carefully, writing down all your working. Don't just plug it all straight in your calculator — you'll make mistakes. To check you're doing it right, make sure you can reproduce the y-values they've already given you.

2) Draw the Curve

1) PLOT THE POINTS CAREFULLY, and don't mix up the x and y values.
2) The points should form a COMPLETELY SMOOTH CURVE. If they don't, they're wrong.

NEVER EVER let one point drag your line off in some ridiculous direction. When a graph is generated from an equation, you never get spikes or lumps — only MISTAKES.

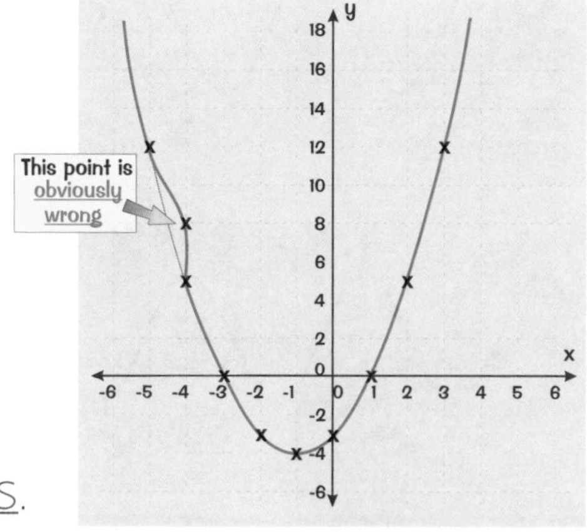

3) Read off the Solutions

Example: "Use your graph to solve the equation $x^2 + 2x - 3 = 0$."

1) Look — the equation you've been asked to solve is what you get when you put $y=0$ into the graph's equation, $y = x^2 + 2x - 3$.
2) To solve the equation, all you do is read the x-values where $y = 0$, i.e. where it crosses the x-axis.
3) So the solutions are $x = -3$ and $x = 1$. (Quadratic eqns usually have 2 solutions.)

The Acid Test:

LEARN THE DETAILS of the method above for DRAWING QUADRATIC GRAPHS and SOLVING THE EQUATION.

Plot the graph of $y = x^2 - x - 6$ (use x-values from -4 to 5).
Use your graph to solve the equation $x^2 - x - 6 = 0$.

The Graphs of Sin, Cos and Tan

You are expected to know these graphs and be able to sketch them from memory.
It really isn't that difficult — the secret is to notice their similarities and differences:

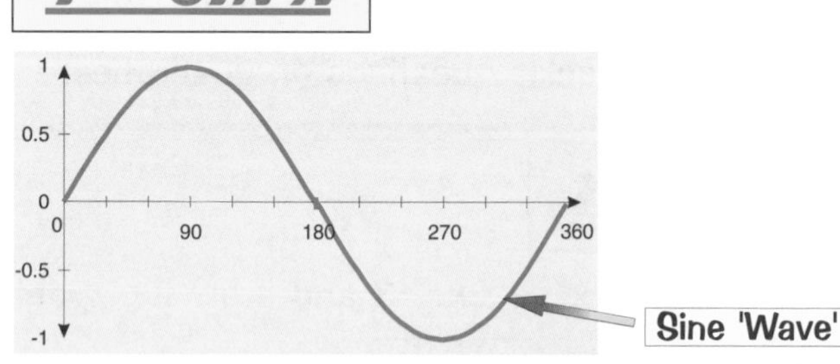

Y = COS X

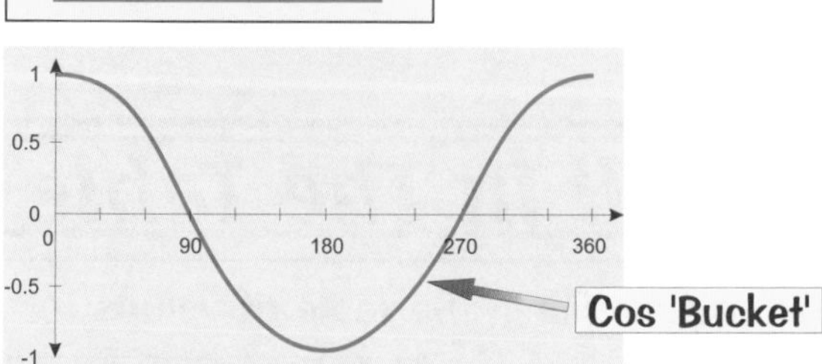

1) For 0° – 360°, the shapes you get are a SINE "WAVE" (One peak, one trough) and a COS "BUCKET" (Starts at the top, dips, and finishes at the top).
2) The underlying shape of both the SIN and COS graphs are identical, (as shown below) when you extend them (indefinitely) in both directions:

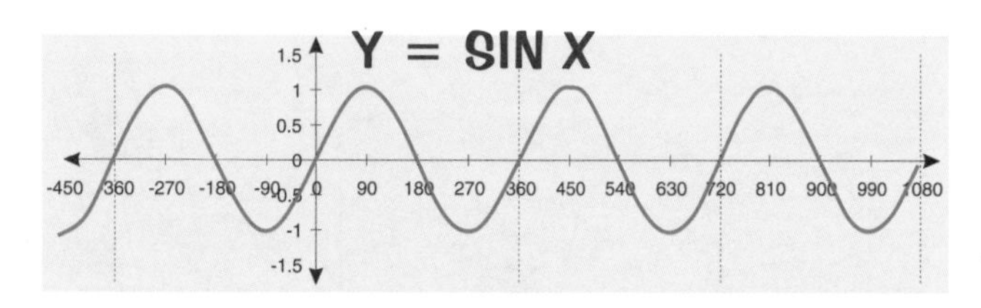

Y = COS X

3) The only difference is that the SIN graph is shifted by 90° → compared to the COS graph.
4) Note that both graphs wiggle between y-limits of exactly +1 and -1.
5) The key to drawing the extended graphs is to first draw the 0 – 360° cycle of either the SIN "WAVE" or the COS "BUCKET" and then repeat it in both directions as shown.

Y = TAN X

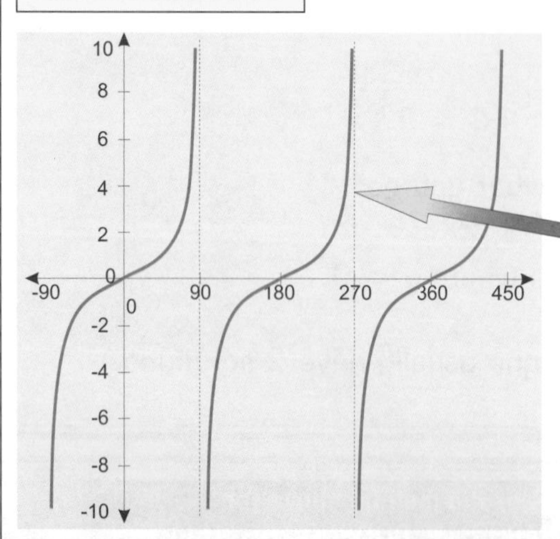

1) The TAN graph BEARS NO RESEMBLANCE to the other two.
2) It behaves in a fairly bizarre way at 90°, 270° etc. ....by disappearing up to + infinity and then reappearing from - infinity on the other side of the asymptote (— a dotted line that the graph never quite touches).
3) So unlike the SIN and COS graphs, Y = TAN X is not limited to values between +1 and -1.
4) You'll also notice that whilst SIN and COS repeat every 360°, the TAN graph repeats every 180°.

The Acid Test: LEARN The FIVE graphs above. Then turn over and draw all five again in full detail.

Angles of Any Size

You can only do this if you've learnt the graphs on the other page:

SIN, COS and TAN for Angles of Any Size

There is ONE BASIC IDEA involved here:

> If you draw a horizontal line at a given value for sin x then it will pick out an infinite number of angles on the x-axis which all have the same value for sin x.

Example 1:

"Find 6 different angles x such that sin x = 0.94"

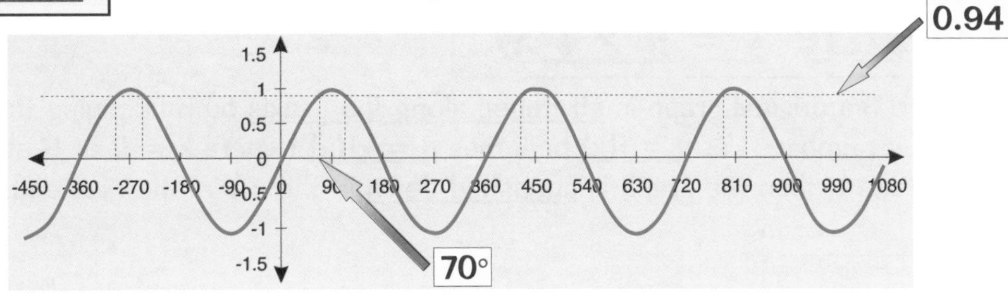

Method

1) Sketch the extended sin x graph.
2) Put a horizontal line across at 0.94.
3) Draw lines down to the x-axis wherever the horizontal crosses the curve.
4) Use your calculator to find inv sin 0.94, to get the first angle (70° in this case).
5) The symmetry is surely obvious. You can see that 70° is 20° away from the peak, so all the other angles are clearly 20° either side of the peaks at 90°, 450°, etc.

> Hence we can say that sin x = +0.94 for all the following angles:
> ..... -290°, -250°, 70°, 110°, 430°, 470°, 790°, 830°....

Example 2:

"Find three other angles which have the same Cosine as 65°."

ANSWER: 1) Use the calculator to find COS 65° = +0.423
2) Draw the extended COS curve and a horizontal line across at + 0.423
3) Draw the vertical lines from the intersections and use symmetry

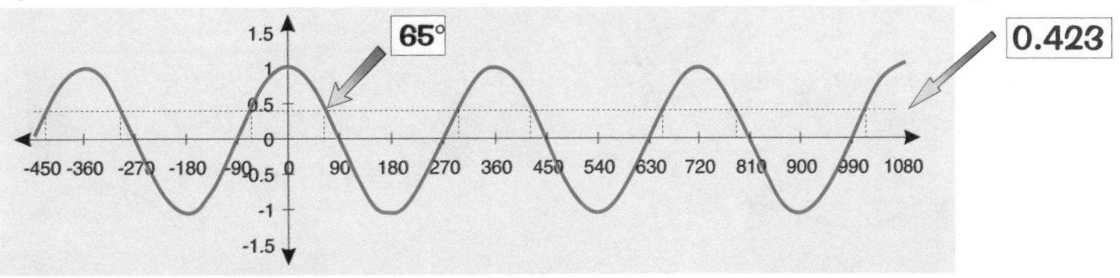

Since 65° is 25° below 90° the other angles shown must be: -425°, -295°, -65°, etc

The Acid Test:

LEARN the method above.
Then turn over and write it all down.

1) Find the first 4 positive values and first two negative values for x such that
a) SIN x = 0.5 b) COS x = -0.67 c) TAN x = 1

Graphs: Shifts and Stretches

Don't be put off by function notation involving f(x). It doesn't mean anything complicated, it's just a fancy way of saying "An equation in x".
In other words "y = f(x)" just means "y = some totally mundane equation in x, which we won't tell you, we'll just call it f(x) instead to see how many of you get in a flap about it".

In a question on transforming graphs they will either use function notation or they'll use a known function instead. There are only four different types of graph transformations so just learn them and be done with it. Here they are in order of difficulty:

1) y-Stretch: $y = k \times f(x)$

This is where the original graph is stretched along the y-axis by multiplying the whole function by a number, i.e. $y = f(x)$ becomes $y = kf(x)$ (where k = 2 or 5 etc.).
If k is less than 1, then the graph is squashed down in the y-direction instead:

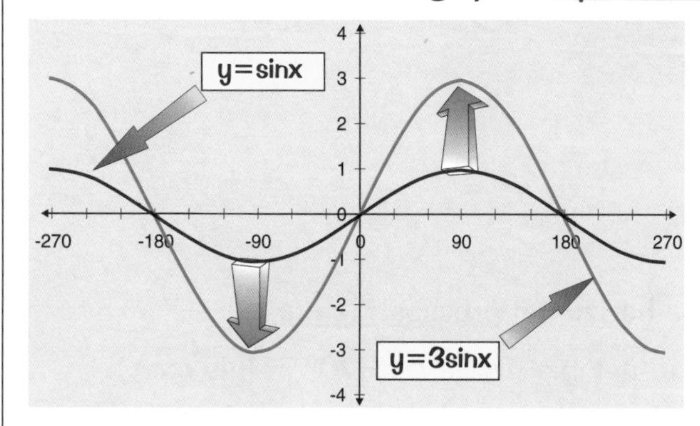

This graph shows $y = f(x)$ and $y = 3f(x)$
(y = SIN X and Y = 3 SIN X)

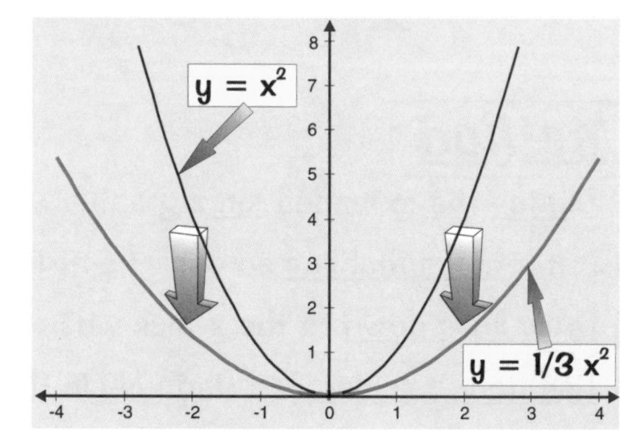

This graph shows $y = f(x)$ and $y = 1/3\ f(x)$
($y = x^2$ and $y = 1/3\ x^2$)

2) y-Shift: $y = f(x) + a$

This is where the whole graph is slid up or down the y-axis with no distortion, and is achieved by simply adding a number onto the end of the equation: $y = f(x) + a$.

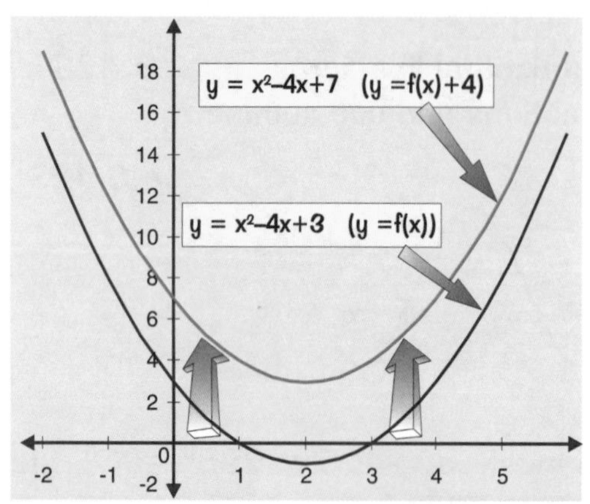

This shows $y = f(x)$ and $y = f(x) + 4$
i.e. $y = x^2 - 4x + 3$, and
$y = (x^2 - 4x + 3) + 4$
or $y = x^2 - 4x + 7$

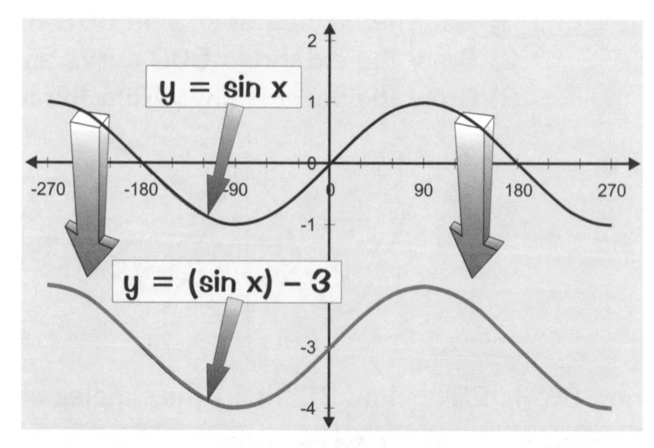

This shows $y = f(x)$ and $y = f(x) - 3$
i.e $y = \sin x$ and $y = (\sin x) - 3$

Graphs: Shifts and Stretches

3) x-Shift: $Y = f(x - a)$

This is where the whole graph slides to the left or right and it only happens when you replace "x" everywhere in the equation with "x – a". These are a bit tricky because they go "the wrong way". In other words if you want to go from $y = f(x)$ to $y = f(x - a)$ you must move the whole graph a distance "a" in the positive x-direction → (and vice versa).

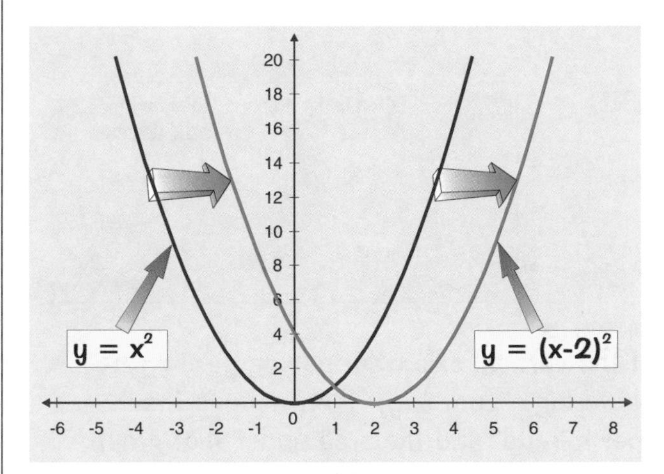

This graph shows $y=f(x)$ and $y=f(x-2)$
(i.e. $y = x^2$ and $y = (x-2)^2$)

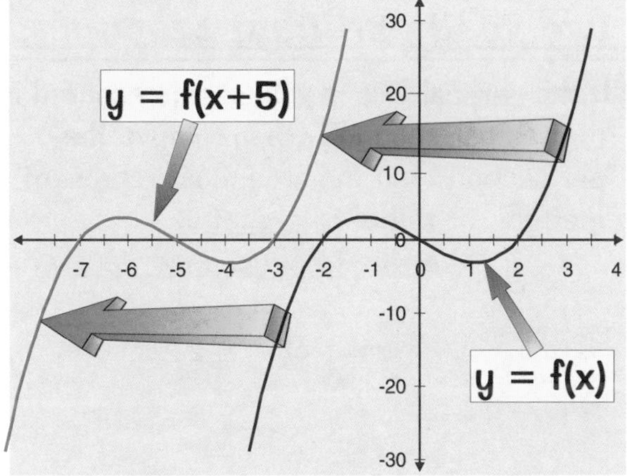

This graph shows $y=f(x)$ and $y=f(x+5)$
i.e. $y=x^3 - 4x$, and $y=(x+5)^3 - 4(x+5)$

4) x-Stretch: $Y = f(kx)$

These go "the wrong way" too — when k is a "multiplier" it scrunches the graph up, whereas when it's a "divider", it stretches the graph out. (The opposite of the y-stretch)

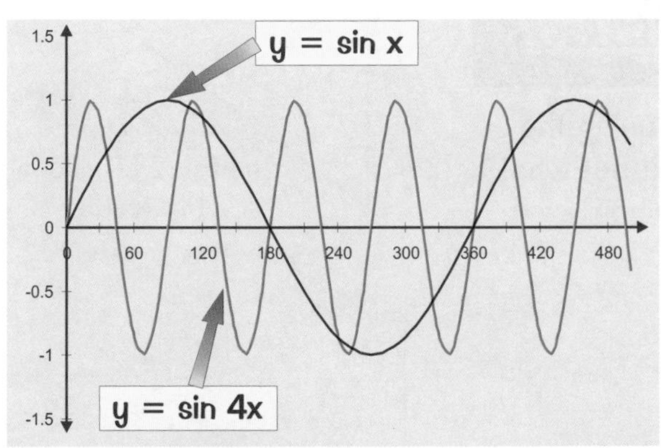

This graph shows
$y = \sin x$ and $y = \sin 4x$
The one that is all squashed up is $y = \sin 4x$. The way to sketch it is simply that with a multiplier of 4, it will be 4 times as squashed up.
(Each full cycle of up-and-down takes ¼ the amount of x-axis as the original graph, so you fit 4 of them into 1 of the other graph)

Remember, if k is a divider, then the graph spreads out. So if the squashed up graph above was the original, $y = f(x)$, then the more spread out one would be $y = f(x/4)$.

The Acid Test:

LEARN the Four types of Graph Transformations, both the effect on the formula and the effect on the graph. Then turn over and draw two examples of each type.

1) Sketch these graphs: a) $y = x^2$ b) $y = x^2 - 4$ c) $y = 3x^2$ d) $y = (x - 3)^2$
e) $y = \cos x$ f) $y = \cos (x+30^0)$ g) $y = \cos x + 3$ h) $y = 2\cos x - 4$

The Meaning of Area and Gradient

The Area Under a Graph Represents the TOTAL SOMETHING

The secret here is the unit of the vertical axis which will be a RATE of some sort, i.e. "SOMETHING per ANOTHER-THING".
To find the meaning of the area under the graph, all you do is remove the "PER ANOTHER-THING" from those units and the "SOMETHING" that's left is what the area under the graph represents. The area is the "TOTAL SOMETHING" in fact.

Two Really Brilliant Examples:

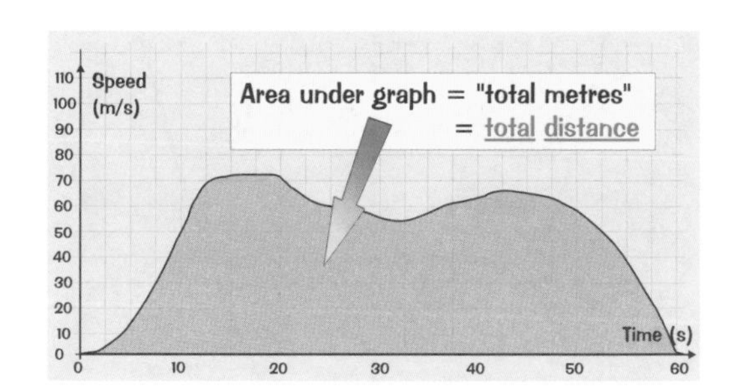

1) If the vertical axis is velocity, (measured in "metres per second"), then remove the "per second" and the area must represent "metres", or rather total metres, i.e. the total distance travelled.

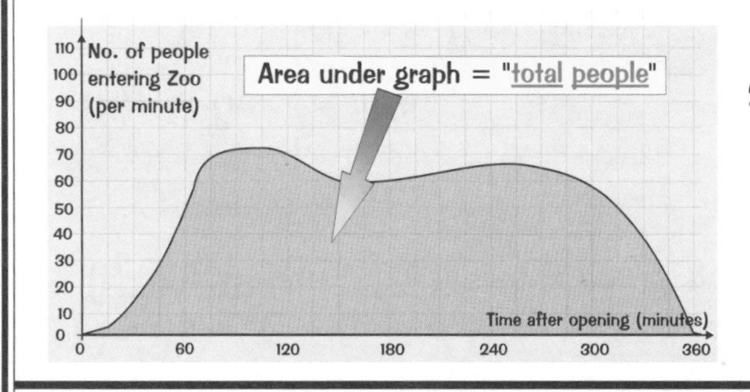

2) If the vertical axis was "people per minute" (entering a zoo, say), then remove the "per minute" and the area under that graph would simply be total people.

(Easy peasy, don't you think?)

The Gradient of a Graph Represents the Rate

No matter what the graph may be, the meaning of the gradient is always simply:

(y-axis UNITS) PER (x-axis UNITS)

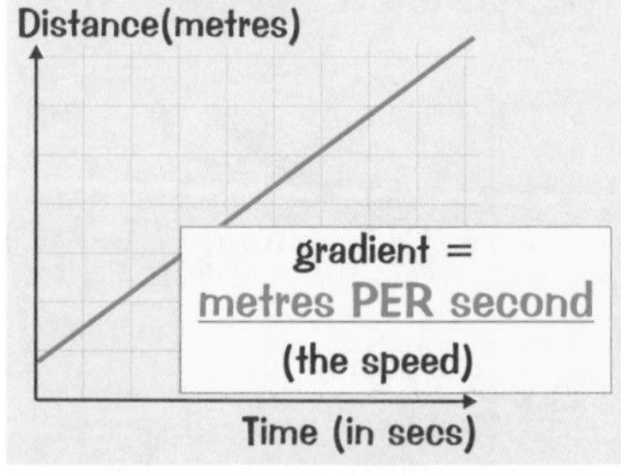

Once you've written "something PER something" using the x- and y-axis UNITS, it's then pretty easy to work out what the gradient represents, as these four examples show.

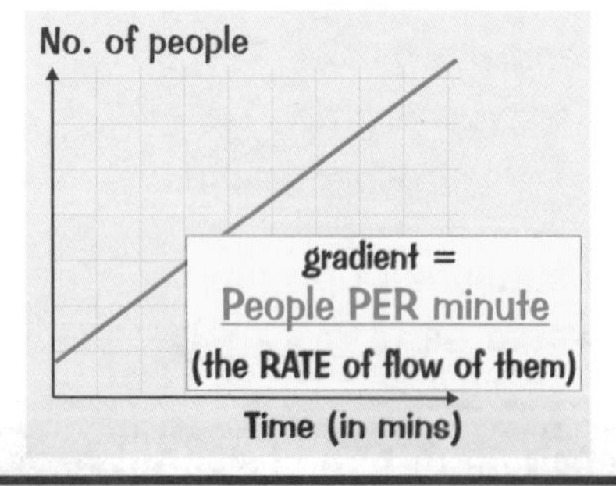

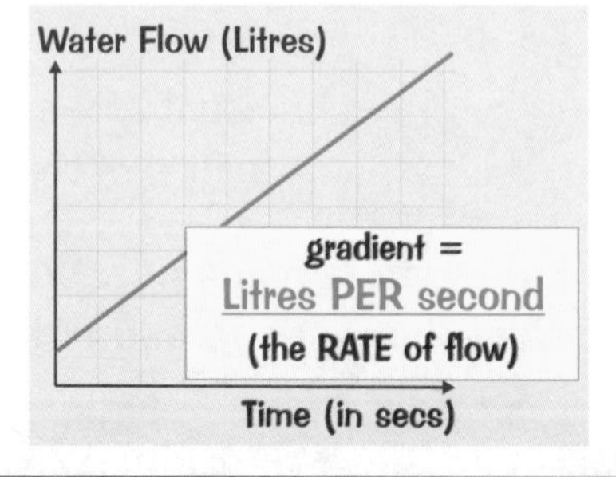

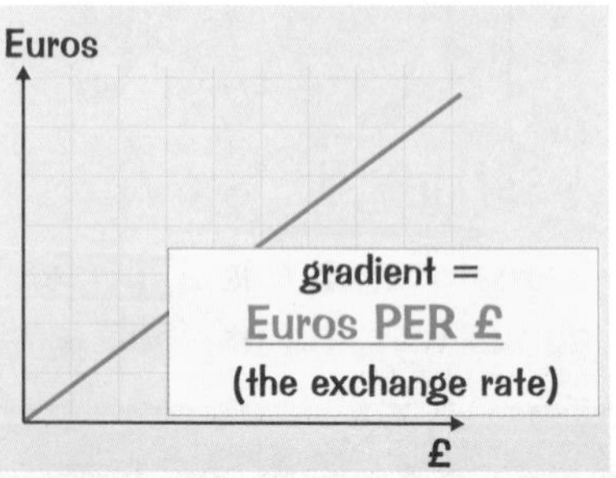

The Acid Test:

LEARN both sections on this page, then turn over and write down the main details from memory.

1) On a graph of "babies born per minute" vs "time in minutes", what would the area represent?
2) If I drew a graph of "miles covered" up the y-axis and "gallons used" along the x-axis, and worked out the gradient, what would the value of it tell me?

Finding Equations From Graphs

The basic idea here is to get the equation for a given curve and there are four main types of equation/curve which you're likely to get in the Exam:

SQUARED FUNCTION:	"$y = ax^2 + b$"	CUBIC FUNCTION:	"$y = ax^3 + b$"
EXPONENTIAL FUNCTION:	"$y = pq^x$"	TRIG FUNCTION:	"$y = D(\sin X) + E$"

It can all seem quite tricky to the uninitiated but once you've cottoned on to the method it's really pretty simple. It's the same easy method for all of them and this is it:

Method

1) You have to find TWO unknowns in the equation (e.g. a and b, or p and q, etc.), which means you'll need TWO pairs of x and y values to stick in the equation.
2) You find these simply by taking the coordinates of two points on the graph.
3) You should always try to take points that lie on either the x-axis or y-axis.
(this makes one of the coordinates ZERO which makes the equations much easier to solve)

Example

The graph below has been obtained from experimental data and the curve appears to be of the form "$H = at^2 + b$". Use the graph to find values for the constants "a" and "b".

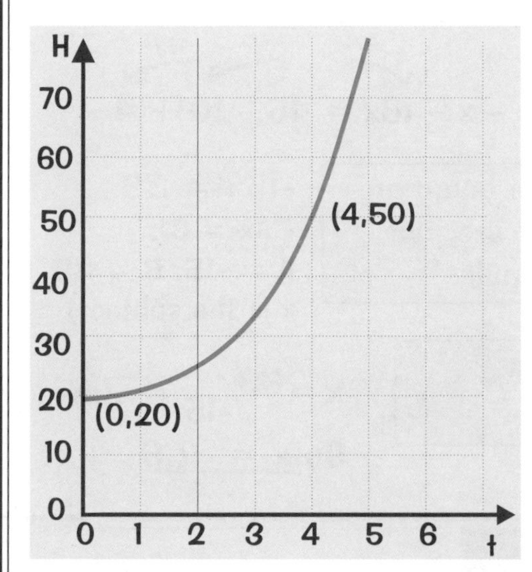

ANSWER:
We can choose any two points on the graph but the most obvious and sensible choices are the two indicated: (0,20) and (4,50).
The best of these of course is (0,20), and sticking these values for H and t in the equation gives:
$20 = 0 + b$ so straight away we know $b=20$

Now using (4,50), together with b=20 gives:
$50 = a \times 16 + 20$
which gives: $a = (50-20)/16 = 1.875$ so $a = 1.9$ (to 2sf)

Hence the equation is $H = 1.9t^2 + 20$

If they give you one of the other equations, then the algebra will be a bit different. However, the basic method is always exactly the same so make sure you know it!

The Acid Test:

LEARN the three points of the method above.
Then turn the page and write them down.

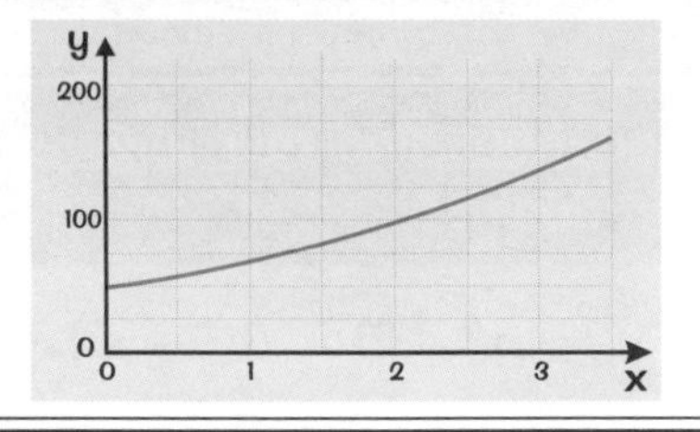

1) The graph shown here is thought to be represented by the equation $y = PQ^x$ where P and Q are unknown constants. Use the graph to find the values of P and Q.

Solving Equations

Solving Equations means finding the value of x from something like: $3x + 5 = 4 - 5x$.
Now, not a lot of people know this, but exactly the same method applies to both solving equations and rearranging formulas, as illustrated on these two pages.

1) EXACTLY THE SAME METHOD APPLIES TO BOTH FORMULAS AND EQUATIONS.
2) THE SAME SEQUENCE OF STEPS APPLIES EVERY TIME.

To illustrate the sequence of steps we'll use this equation: $\sqrt{2 - \frac{x+4}{2x+5}} = 3$

The Six Steps Applied to Equations

1) Get rid of any square root signs by squaring both sides: $2 - \frac{x+4}{2x+5} = 9$

2) Get everything off the bottom by cross-multiplying up to EVERY OTHER TERM:

$$2 - \frac{x+4}{2x+5} = 9 \quad \Rightarrow \quad 2(2x+5) - (x+4) = 9(2x+5)$$

3) Multiply out any brackets: $4x + 10 - x - 4 = 18x + 45$

4) Collect all subject terms on one side of the "=" and all non-subject terms on the other. Remember to reverse the +/– sign of any term that crosses the "="

+18x moves across the "=" and becomes -18x
+10 moves across the "=" and becomes -10
-4 moves across the "=" and becomes +4

$$4x - x - 18x = 45 - 10 + 4$$

5) Combine together like terms on each side of the equation, and reduce it to the form "Ax = B", where A and B are just numbers (or bunches of letters, in the case of formulas):

$-15x = 39$
("Ax = B": A = -15, B = 39, x is the subject)

6) Finally slide the A underneath the B to give "$x = B/A$", divide, and that's your answer.

$x = 39/-15 = -2.6$
So $x = -2.6$

The Seventh Step (if You Need It)

If the term you're trying to find is squared, don't panic.

Follow steps 1) to 6) like normal, but solve it for, x^2 instead of x: $x^2 = 9$

7) Take the square root of both sides and stick a ± sign in front of the expression on the right: $x = \pm 3$

Don't forget the ± sign... (P.31 if you don't know what I mean).

The Acid Test:

LEARN the 7 STEPS for solving equations and rearranging formulas. Turn over and write them down.

1) Solve these equations: a) $5(x + 2) = 8 + 4(5 - x)$ b) $\frac{4}{x+3} = \frac{6}{4-x}$ c) $x^2 - 21 = 3(5 - x^2)$

Rearranging Formulas

Rearranging Formulas means making one letter the subject, e.g. getting "y= " from something like $2x + z = 3(y + 2p)$.
Generally speaking "solving equations" is easier, but don't forget:

1) EXACTLY THE SAME METHOD APPLIES TO BOTH FORMULAS AND EQUATIONS.
2) THE SAME SEQUENCE OF STEPS APPLIES EVERY TIME.

We'll illustrate this by making "y" the subject of this formula: $M = \sqrt{2K - \frac{K^2}{2y+1}}$

The Six Steps Applied to Formulas

1) Get rid of any square root signs by squaring both sides: $M^2 = 2K - \frac{K^2}{2y+1}$

2) Get everything off the bottom by cross-multiplying up to EVERY OTHER TERM:

$$M^2 = 2K - \frac{K^2}{2y+1} \Rightarrow M^2(2y+1) = 2K(2y+1) - K^2$$

3) Multiply out any brackets: $2yM^2 + M^2 = 4Ky + 2K - K^2$

4) Collect all subject terms on one side of the "=" and all non-subject terms on the other. Remember to reverse the +/– sign of any term that crosses the "="

+4Ky moves across the "=" and becomes –4Ky
+M² moves across the "=" and becomes –M²

$$2yM^2 - 4Ky = -M^2 + 2K - K^2$$

5) Combine together like terms on each side of the equation, and reduce it to the form "Ax = B", where A and B are just bunches of letters which DON'T include the subject (y). Note that the LHS has to be FACTORISED:

$$(2M^2 - 4K)y = 2K - K^2 - M^2$$

("Ax = B" i.e. $A = (2M^2 - 4K)$, $B = 2K - K^2 - M^2$, y is the subject)

6) Finally slide the A underneath the B to give "$x = B/A$", (cancel if possible) and that's your answer. So $y = \frac{2K - K^2 - M^2}{(2M^2 - 4K)}$

The Seventh Step (if You Need It)

If the term you're trying to make the subject of the equation is squared, this is what you do: $M = \sqrt{2K - \frac{K^2}{2y^2+1}}$

Follow steps 1) to 6), $y^2 = \frac{2K - K^2 - M^2}{(2M^2 - 4K)}$ and then...

(I've skipped steps 1) - 6) because they're exactly the same as the first example — but with y² instead of y.)

7) Take the square root of both sides and stick a ± sign in front of the expression on the right: $y = \pm\sqrt{\frac{2K - K^2 - M^2}{(2M^2 - 4K)}}$

Remember — square roots can be +ve or –ve. See P.31.

The Acid Test:

LEARN the 7 STEPS for solving equations and rearranging formulas. Turn over and write them down.

1) Rearrange " $F = \frac{9}{5}C + 32$ " from "F= ", to "C= " and then back the other way.

2) Make p the subject of these: a) $\frac{p}{p+y} = 4$ b) $\frac{1}{p} = \frac{1}{q} + \frac{1}{r}$ c) $\frac{1}{p^2} = \frac{1}{q} + \frac{1}{r}$

Inequalities

Inequalities aren't half as difficult as they look. Once you've learned the tricks involved, most of the algebra for them is identical to ordinary equations.

THE INEQUALITY SYMBOLS:

$>$ means "Greater than" $\geqslant$ means "Greater than or equal to

$<$ means "Less than" $\leqslant$ means "Less than or equal to"

REMEMBER, the one at the BIG end is BIGGEST

so $x > 4$ and $4 < x$ both say: "x is greater than 4"

Algebra With Inequalities

$5x < x + 2$

$5x = x + 2$

The thing to remember here is that inequalities are just like regular equations in the sense that all the normal rules of algebra apply WITH ONE BIG EXCEPTION:

Whenever you MULTIPLY OR DIVIDE BY A NEGATIVE NUMBER, you must FLIP THE INEQUALITY SIGN.

Three Important Examples

1) Solve $5x < 6x + 2$

The equivalent equation is $5x = 6x + 2$, which is easy — and so is the inequality:

First subtract 6x : $5x - 6x < 2$ which gives $-x < 2$

Then divide both sides by -1: $x > -2$ (i.e. x is greater than -2)

(NOTE: The < has flipped around into a >, because we divided by a –ve number)

2) Find all integer values of x where $4 \leq x < 1$

This type of expression is very common — you must learn them in this way:

" x is between -4 and +1, possibly equal to -4 but never equal to +1 ".

(Obviously the answers are -4, -3, -2, -1, 0 (but not 1))

3) Find the range of values of x where $x^2 \leq 25$

The trick here is: Don't forget the negative values.

Square-rooting both sides gives $x \leq 5$. However, this is only half the story, because $-5 \leq x$ is also true. There is little alternative but to simply learn this:

1) $x^2 \leqslant 25$ gives the solution $-5 \leqslant x \leqslant 5$,
(x is between -5 and 5, possibly equal to either)

2) $x^2 \geqslant 36$ gives the solution: $x \leqslant -6$ or $6 \leqslant x$
(x is "less than or equal to -6" or "greater than or equal to +6")

The Acid Test:

LEARN all of this page including the Three important Examples, then turn over and write it all down.

1) Solve this inequality: $4x + 3 \leq 6x + 7$

2) Find all integer values p, such that a) $p^2 < 49$ b) $-20 < 4p \leq 17$

Graphical Inequalities

These questions always involve <u>shading a region on a graph</u>. The method sounds very complicated, but once you've seen it in action with an example, you'll see it's OK...

Method

1) **<u>CONVERT each INEQUALITY to an EQUATION</u>**

   by simply putting an "=" in place of the "<" or ">"

2) **<u>DRAW THE GRAPH FOR EACH EQUATION</u>**

3) **<u>Work out WHICH SIDE of each line you want</u>**

   Put x=0 and y=0 into the inequality to see if the <u>ORIGIN</u> is on the correct side.

4) **<u>SHADE THE REGION this gives you</u>**

Example

"Shade the region represented by:
$x + y < 5$, $y > x + 2$ and $y > 1$"

1) <u>CONVERT EACH INEQUALITY TO AN EQUATION</u>:
   The inequalities become $y = x + 2$, $x + y = 5$ and $y = 1$

2) <u>DRAW THE GRAPH FOR EACH EQUATION</u> (see p41 or p76)

3) <u>WORK OUT WHICH SIDE OF EACH LINE YOU WANT</u>
   This is the fiddly bit. Substitute $x = 0$ and $y = 0$ (the origin) into each inequality and see if this makes the inequality <u>true</u> or <u>false</u>.

   <u>In $x + y < 5$:</u>
   $x = 0$, $y = 0$ gives $0 < 5$ which is <u>true</u>.
   This means the <u>origin</u> is on the <u>correct</u> side of the line.

   <u>In $y > x + 2$:</u>
   $x = 0$, $y = 0$ gives $0 > 2$ which is <u>false</u>.
   So the origin is on the <u>wrong side</u> of this line.

   <u>In $y > 1$:</u>
   $x = 0$, $y = 0$ gives $0 > 1$ which is <u>false</u>.
   So the origin is on the <u>wrong side</u> of this line too.

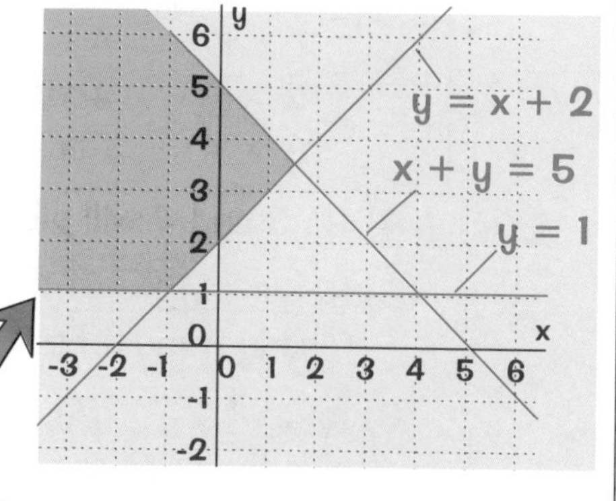

4) <u>SHADE THE REGION</u>
   You want the region that satisfies all of these:
   – below $x + y = 5$ (because the origin <u>is</u> on this side)
   – left of $y = x + 2$ (because the origin <u>isn't</u> on this side)
   – above $y = 1$ (because the origin <u>isn't</u> on this side).

The Acid Test:

LEARN the <u>Four Steps</u> for doing <u>graphical inequalities</u>, then <u>turn over</u> and <u>write them down</u>.

1) Show on a graph the region described by the following three conditions:
   $x + y < 6$, $y > 0.5$, $y < 2x - 2$

Factorising Quadratics

There are several ways of solving a quadratic equation as detailed on the following pages. You need to know all the methods.

Factorising a Quadratic

"Factorising a quadratic" means "putting it into 2 brackets" — you'll need to remember that.
(There are several different methods for doing this, so stick with the one you're happiest with. If you have no preference then learn the one below.)

The standard format for quadratic equations is: $ax^2 + bx + c = 0$
Most Exam questions have $a = 1$, making them much easier.

E.g. $x^2 + 3x + 2 = 0$ (See next page for when a is not 1)

Factorising Method When a = 1

1) ALWAYS rearrange into the STANDARD FORMAT: $ax^2 + bx + c = 0$
2) Write down the TWO BRACKETS with the x's in: $(x \quad)(x \quad)=0$
3) Then find 2 numbers that MULTIPLY to give "c" (the end number) but also ADD/SUBTRACT to give "b" (the coefficient of x)
4) Put them in and check that the +/– signs work out properly.

An Example

"Solve $x^2 - x = 12$ by factorising."

ANSWER: 1) First rearrange it (into the standard format): $x^2 - x - 12 = 0$

2) a=1, so the initial brackets are (as ever): $(x \quad)(x \quad)=0$

3) We now want to look at all pairs of numbers that multiply to give "c" (=12), but which also add or subtract to give the value of b:

1×12	Add/subtract to give:	13 or 11	
2×6	Add/subtract to give:	8 or 4	this is what we're
3×4	Add/subtract to give:	7 or (1)	← after ($=\pm b$)

4) So 3 and 4 will give $b = \pm 1$, so put them in: $(x \quad 3)(x \quad 4)=0$

5) Now fill in the +/– signs so that the 3 and 4 add/subtract to give -1 (=b), Clearly it must be +3 and –4 so we'll have: $(x + 3)(x - 4)=0$

6) As an ESSENTIAL check, EXPAND the brackets out again to make sure they give the original equation:
$(x + 3)(x - 4) = x^2 + 3x - 4x - 12 = x^2 - x - 12$

We're not finished yet mind, because $(x + 3)(x - 4)=0$ is only the factorised form of the equation — we have yet to give the actual SOLUTIONS. This is very easy:

7) THE SOLUTIONS are simply the two numbers in the brackets, but with OPPOSITE +/– SIGNS: i.e. $x = -3$ or $+4$

Make sure you remember that last step. It's the difference between SOLVING THE EQUATION and merely factorising it.

Factorising Quadratics

When "a" is not 1

E.g. $3x^2 + 5x + 2 = 0$

The basic method is still the same but it's a lot messier. Chances are, the Exam question will be with a=1, so make sure you can do that type easily. Only then should you try to get to grips with these harder ones.

An Example

"Solve $3x^2 + 7x = 6$ by factorising."

1) First rearrange it (into the standard format): $3x^2 + 7x - 6 = 0$

2) Now because a = 3, the two x-terms in the brackets will have to multiply to give $3x^2$ so the initial brackets will have to be: $(3x \quad)(x \quad)=0$

(i.e. you put in the x-terms first, with coeffts. that will multiply to give "a")

3) We now want to look at all pairs of numbers that multiply with each other to give "c" (=6, ignoring the minus sign for now): i.e. 1×6 and 2×3

4) Now the difficult bit: to find the combination which does this:

multiply with the 3x and x terms in the brackets and then add or subtract to give the value of b (=7):

The best way to do this is by trying out all the possibilities in the brackets until you find the combination that works. Don't forget that EACH PAIR of numbers can be tried in TWO different positions:

(3x 1)(x 6) multiplies to give 18x and 1x which add/subtract to give 17x or 19x
(3x 6)(x 1) multiplies to give 3x and 6x which add/subtract to give 9x or 3x
(3x 3)(x 2) multiplies to give 6x and 3x which add/subtract to give 9x or 3x
(3x 2)(x 3) multiplies to give 9x and 2x which add/subtract to give 11x or 7x

4) So (3x 2)(x 3) is the combination that gives b = 7, (give or take a +/–)

5) Now fill in the +/– signs so that the combination will add/subtract to give +7 (=b). Clearly it must be +3 and –2 which gives rise to +9x and -2x
So the final brackets are: $(3x - 2)(x + 3)$

6) As an ESSENTIAL check, EXPAND the brackets out again to make sure they give the original equation:
$(3x - 2)(x + 3) = 3x^2 + 9x - 2x - 6 = 3x^2 + 7x - 6$

7) The last step is to get THE SOLUTIONS TO THE EQUATION: $(3x - 2)(x + 3)=0$ which you do by separately putting each bracket = 0 :

i.e. $(3x - 2)=0 \Rightarrow x = 2/3$ $\quad (x + 3)=0 \Rightarrow x = -3$

Don't forget that last step. Again, it's the difference between SOLVING THE EQUATION and merely factorising it.

The Acid Test:

LEARN the 7 steps for solving quadratics by factorising, both for "a = 1" and "a ≠ 1".

1) Solve these by factorising: a) $x^2 + 5x - 24 = 0$ b) $x^2 - 6x + 9 = 16$
c) $(x + 3)^2 - 3 = 13$ d) $5x^2 - 17x - 12 = 0$

The Quadratic Formula

The solutions to any quadratic equation $ax^2 + bx + c = 0$ are given by this formula:

$$x = \frac{-b \pm \sqrt{b^2 - 4ac}}{2a}$$

LEARN THIS FORMULA — If you can't learn it, there's no way you'll be able to use it in the Exam, even if they give it to you. Using it should, in principle, be quite straightforward. As it turns out though there are quite a few pitfalls, so TAKE HEED of these crucial details:

Using The Quadratic Formula

1) Always write it down in stages as you go. Take it nice and slowly — any fool can rush it and get it wrong, but there's no marks for being a clot.
2) MINUS SIGNS. Throughout the whole of algebra, minus signs cause untold misery because people keep forgetting them. In this formula, there are two minus signs that people keep forgetting: the $-b$ and the $-4ac$.

   The $-4ac$ causes particular problems when either "a" or "c" is negative, because it makes the $-4ac$ effectively $+4ac$ — so learn to spot it as a HAZARD before it happens.

   WHENEVER YOU GET A MINUS SIGN, THE ALARM BELLS SHOULD ALWAYS RING!
3) Remember you divide ALL of the top line by 2a, not just half of it.
4) Don't forget it's 2a on the bottom line, not just a. This is another common mistake.

EXAMPLE:

"Find the solutions of $3x^2 + 7x = 1$ to 2 decimal places."

(The mention of decimal places in exam questions is a very big clue to use the formula rather than trying to factorise it!)

METHOD:

1) First get it into the form $ax^2 + bx + c = 0$: $3x^2 + 7x - 1 = 0$
2) Then carefully identify a, b and c: $a = 3, \quad b = 7, \quad c = -1$
3) Put these values into the quadratic formula and write down each stage:

$$x = \frac{-b \pm \sqrt{b^2 - 4ac}}{2a} = \frac{-7 \pm \sqrt{7^2 - 4 \times 3 \times -1}}{2 \times 3} = \frac{-7 \pm \sqrt{49 + 12}}{6}$$

$$= \frac{-7 \pm \sqrt{61}}{6} = \frac{-7 \pm 7.81}{6} = 0.1350 \text{ or } -2.468$$

So to 2 DP, the solutions are: $x = 0.14$ or -2.47

4) Finally as a check put these values back into the original equation:
   E.g. for $x = 0.1350$: $3 \times 0.135^2 + 7 \times 0.135 = 0.999675$,
   which is 1, as near as ...

The Acid Test:

LEARN the 4 CRUCIAL DETAILS and the 4 STEPS OF THE METHOD for using the Quadratic Formula, then TURN OVER AND WRITE THEM ALL DOWN.

1) Find the solutions of these equations (to 2 DP) using the Quadratic formula:
   a) $x^2 + 10x - 4 = 0$ b) $3x^2 - 3x = 2$ c) $(2x + 3)^2 = 15$

Completing the Square

$$x^2 + 12x - 5 = (x + 6)^2 - 41$$

The SQUARE... ...COMPLETED

Solving Quadratics by "Completing The Square"

This is quite a clever way of solving quadratics, but is perhaps a bit confusing at first. The name "Completing the Square" doesn't help — it's called that because of the method where you basically:

1) write down a SQUARED bracket, and then
2) stick a number on the end to "COMPLETE" it.

It's quite easy if you learn all the steps — some of them aren't all that obvious.

Method:

1) As always, REARRANGE THE QUADRATIC INTO THE STANDARD FORMAT:
$$ax^2 + bx + c = 0$$
2) If "a" is not 1 then divide the whole equation by "a" to make sure it is!
3) Now WRITE OUT THE INITIAL BRACKET: $(x + b/2)^2$

   NB: THE NUMBER IN THE BRACKET is always HALF THE (NEW) VALUE OF "b"
4) MULTIPLY OUT THE BRACKETS and COMPARE TO THE ORIGINAL to find what extra is needed, and add or subtract the adjusting amount.

Example:

"Express $x^2 - 6x - 7 = 0$ as a completed square, and hence solve it."

The equation is already in the standard form and "a" = 1, so:

1) The coefficient of x is -6, so the squared brackets must be: $(x - 3)^2$
2) Square out the brackets: $x^2 - 6x + 9$, and compare to the original: $x^2 - 6x - 7$.
   To make it like the original equation it needs -16 on the end, hence we get:

$(x - 3)^2 - 16 = 0$ as the alternative version of $x^2 - 6x - 7 = 0$

Don't forget though, we wish to SOLVE this equation, which entails these 3 special steps:

1) Take the 16 over to get: $(x - 3)^2 = 16$.
2) Then SQUARE ROOT BOTH SIDES: $(x - 3) = \pm 4$ AND DON'T FORGET THE $\pm$
3) Take the 3 over to get: $x = \pm 4 + 3$ so $x = 7$ or -1 (don't forget the $\pm$)

The Acid Test:

LEARN the 4 STEPS OF THE METHOD for completing the square and the 3 SPECIAL STEPS for SOLVING THE EQUATION you get from it.

1) Now turn over and write it all down to see what you've learned. (Frightening isn't it)
2) Find the solutions of these equations (to 2 DP) by completing the square:
   a) $x^2 + 10x - 4 = 0$ b) $3x^2 - 3x = 2$ c) $(2x + 3)^2 = 15$

Trial and Improvement

In principle, this is an easy way to find approximate answers to quite complicated equations. BUT......you have to make an effort to LEARN THE FINER DETAILS of this method, otherwise you'll never get the hang of it.

Method

1) **SUBSTITUTE TWO INITIAL VALUES into the equation that give OPPOSITE CASES.** *These are usually suggested in the question. If not, you'll have to think of your own. "Opposite cases" means one answer too big, one too small. If your values don't give opposite cases try again.*

2) **Now CHOOSE YOUR NEXT VALUE IN BETWEEN THE PREVIOUS TWO, and SUBSTITUTE it into the equation.** *Continue this process, always choosing a new value between the two closest opposite cases, (and preferably nearer to the one which is closest to the answer you want).*

3) **AFTER ONLY 3 OR 4 STEPS you should have 2 numbers which are to the right degree of accuracy but DIFFER BY 1 IN THE LAST DIGIT.** *For example if you had to get your answer to 2 DP then you'd eventually end up with say 5.43 and 5.44, with these giving OPPOSITE results of course.*

4) **At this point you ALWAYS take the Exact Middle Value to decide which is the answer you want.** *E.g. for 5.43 and 5.44, you'd try 5.435 to see if the real answer was between 5.43 and 5.435 or between 5.435 and 5.44 (See below).*

Example:

The equation $x^2 + x = 14$ has a solution between 3 and 3.5. Find this solution to 1 DP.

Try $x = 3$	$3^2 + 3 = 12$	(Too small)
Try $x = 3.5$	$3.5^2 + 3.5 = 15.75$	(Too big)

← (2 opposite cases)

14 is what we want and it's slightly closer to 15.75 than it is to 12 so we'll choose our next value for x a bit closer to 3.5 than 3

Try $x = 3.3$ $\quad 3.3^2 + 3.3 = 14.19$ (Too big)

Good, this is very close, but we need to see if 3.2 is still too big or too small:

Try $x = 3.2$ $\quad 3.2^2 + 3.2 = 13.44$ (Too small)

Good, now we know that the answer must be between 3.2 and 3.3. To Find out which one it's nearest to, we have to try the EXACT MIDDLE VALUE: 3.25

Try $x = 3.25$ $\quad 3.25^2 + 3.25 = 13.81$ (Too small)

This tells us with certainty that the solution must be between 3.25 (too small) and 3.3 (too big), and so to 1 DP it must round up to 3.3. ANSWER = 3.3

The Acid Test:

"LEARN and TURN" — If you don't actually commit it to memory, then you've wasted your time even reading it.

To succeed with this method you must LEARN the 4 steps above. Do it now, and practise until you can write them down without having to look back at them. It's not as difficult as you think.

1) The equation $x^2 - 2x = 1$ has a solution between 2 and 3. Find it to 1 DP.

Simultaneous Equations and Graphs

When you have two graphs which represent two separate equations, there are two ways the question can present it: two simultaneous equations or a single merged equation.
In either case the solutions will simply be where the two graphs cross...

1) Two Graphs and Two Separate Equations

Example 1: "Draw the graphs for "$y = 2x + 3$" and "$y = 6 - 4x$" and then use your graphs to solve them."

Just draw the graphs and read off the x- and y- values where they cross...

$x = ½$ and $y = 4$

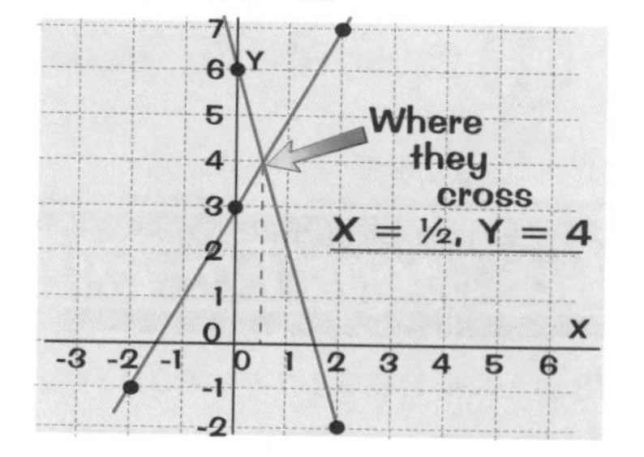

Example 2: "By drawing graphs, solve the simultaneous equations $x^2 + y^2 = 16$ and $y = 2x + 1$."

1) DRAW BOTH GRAPHS.
$x^2 + y^2 = 16$ is the equation of a circle, centre (0, 0), radius 4 (see p.78)

2) LOOK FOR WHERE THE GRAPHS CROSS.
The straight line crosses the circle at two points. Reading the x and y values of these points gives the solutions $x = 1.4$, $y = 3.8$ and $x = -2.2$, $y = -3.4$ (to 1 decimal place).

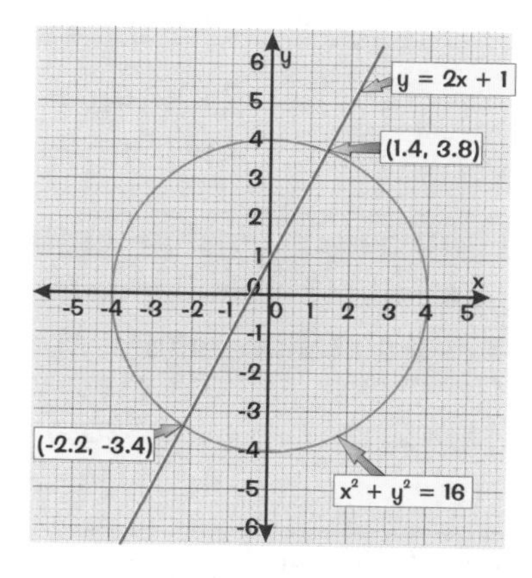

2) Two Graphs but Just ONE Equation, or so it seems...

Example "Using the graphs shown for $y = 4 + ½x$ and $y = 6 - x^2/3$, solve the equation: $x^2/3 + ½x - 2 = 0$."

ANSWER: Learn these important steps:

1) Equating the equations of the two graphs gives this:
$6 - x^2/3 = 4 + ½x$ (a sort of "merged equation")

2) Now bring it all onto one side and you end up with:
$x^2/3 + ½x - 2 = 0$ (the equation in the question!)

3) Hence the solutions to that equation are where the two initial equations ($y = 4 + ½x$ and $y = 6 - x^2/3$) are equal — i.e. where their graphs cross, which as the graph shows is at: $x = 1.8$ or $x = -3.3$.

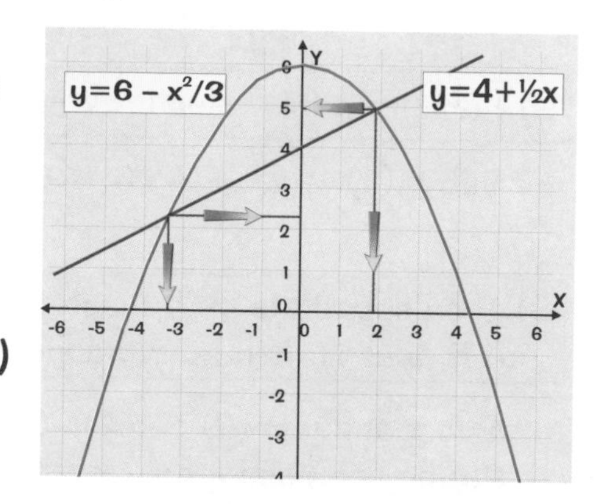

The Acid Test:

LEARN THE IMPORTANT DETAILS on this page, then turn over and write them all down. Keep trying till you can.

1) Use graphs to find the solutions to these pairs of equations:
   a) $y = 4x - 4$ and $y = 6 - x$ b) $y = 2x$ and $y = 6 - 2x$
2) Draw the graphs of $y = 2x^2 - 4$ and $y = 2 - x$ and hence solve $2x^2 + x = 6$.

Simultaneous Equations

You've seen the easy way to solve simultaneous equations using graphs. Now it's time to learn the less fun algebra methods. The rules are really quite simple, but you must follow ALL the steps, in the right order, and treat them as a strict method.

There are two types of simultaneous equations you could get
— EASY ONES (where both equations are linear) and TRICKY ONES (where one's quadratic).

(1) $2x = 6 - 4y$ and $-3 - 3y = 4x$ (2) $7x + y = 1$ and $2x^2 - y = 3$

(1) Six Steps For EASY Simultaneous Equations

We'll use these two equations for our example: $2x = 6 - 4y$ and $-3 - 3y = 4x$

1) Rearrange both equations into the form $ax + by = c$ where a,b,c are numbers, (which can be negative). Also label the two equations —(1) and —(2)

$2x + 4y = 6$ —(1)
$-4x - 3y = 3$ —(2)

2) You need to match up the numbers in front (the "coefficients") of either the x's or y's in both equations. To do this you may need to multiply one or both equations by a suitable number. You should then relabel them: —(3) and —(4)

(1)×2 : $4x + 8y = 12$ —(3)
$-4x - 3y = 3$ —(4)

3) Add or subtract the two equations to eliminate the terms with the same coefficient
If the coefficients are the same (both +ve or both –ve) then SUBTRACT
If the coefficients are opposite (one +ve and one –ve) then ADD

(3) + (4) $0x + 5y = 15$

4) Solve the resulting equation to find whichever letter is left in it.

$5y = 15 \Rightarrow y = 3$

5) Substitute this value back into equation (1) and solve it to find the other quantity.

Sub in (1): $2x + 4\times3 = 6 \Rightarrow 2x + 12 = 6 \Rightarrow 2x = -6 \Rightarrow x = -3$

6) Then substitute both these values into equation (2) to make sure it works.
If it doesn't then you've done something wrong and you'll have to do it all again.

Sub x and y in (2) : $-4\times-3 - 3\times3 = 12 - 9 = 3$, which is right, so it's worked.
So the solutions are: $x = -3$, $y = 3$

The Acid Test:

LEARN the 6 Steps for solving EASY Simultaneous Equations.

1) Remember, you only know them when you can write them all out from memory, so turn over the page and see if you can write down all six steps. Then try again.
2) Then apply the 6 steps to find F and G given that
$2F - 10 = 4G$ and $3G = 4F - 15$

Simultaneous Equations

2 Seven Steps For TRICKY Simultaneous Equations

Example: Solve these two equations simultaneously: $7x + y = 1$ and $2x^2 - y = 3$

1) Rearrange the quadratic equation so that you have the non-quadratic unknown on its own. Label the equations (1) and (2).

$7x + y = 1$ — (1)

$y = 2x^2 - 3$ — (2)

2) Substitute the quadratic expression into the other equation. You'll get another equation — label it (3).

$7x + y = 1$ — (1)

$y = 2x^2 - 3$ — (2)

In this example you just shove the expression for y into equation (1), in place of y.

$7x + (2x^2 - 3) = 1$ — (3)

3) Rearrange to get a quadratic equation. And guess what... You've got to solve it.

Remember — if it won't factorise, you can either use the formula or complete the square. Have a look at P.92-93 for more details.

$2x^2 + 7x - 4 = 0$

That factorises into:
$(2x - 1)(x + 4) = 0$

Check this step by multiplying out again:
$(2x - 1)(x + 4) = 2x^2 - x + 8x - 4 = 2x^2 + 7x - 4$ ☺

So, $2x - 1 = 0$ OR $x + 4 = 0$
In other words, $x = 0.5$ OR $x = -4$

4) Stick the first value back in one of the original equations (pick the easy one).

(1) $7x + y = 1$ Substitute in $x = 0.5$: $3.5 + y = 1$, so $y = 1 - 3.5 = -2.5$

5) Stick the second value back in the same original equation (the easy one again).

(1) $7x + y = 1$ Substitute in $x = -4$: $-28 + y = 1$, so $y = 1 + 28 = 29$

6) Substitute both pairs of answers back into the other original equation to check they work.

(2) $y = 2x^2 - 3$

Substitute in $x = 0.5$ and $y = -2.5$: $-2.5 = (2 \times 0.25) - 3 = -2.5$ — jolly good.

Substitute in $x = -4$ and $y = 29$: $29 = (2 \times 16) - 3 = 29$ — smashing.

7) Write the pairs of answers out again, *CLEARLY*, at the bottom of your working.

The two pairs of answers are: $x = 0.5$ and $y = -2.5$ **OR** $x = -4$ and $y = 29$

(Do this even if you think it's pointless and stupid. If there's even the remotest chance of the examiner getting the pairs mixed up, it's worth a tiny bit of extra effort, don't you think.)

The Acid Test:

LEARN the 7 Steps for solving TRICKY Simultaneous Equations.

Apply the 7 steps to find f and g, given that:

a) $f = g^2 + 4$ and $f - 6g - 4 = 0$
b) $13g - f = -7$ and $3g^2 - f = 3$
c) $4g + f = 3$ and $f = 4g^2$
d) $g = 4f^2 - 3$ and $g + 11f = 0$

Compound Growth and Decay

This can also be called "Exponential" Growth or Decay.

The Formula

This topic is simple if you LEARN THIS FORMULA. If you don't, it's pretty well impossible:

$$N = N_0(1 + r/100)^n$$

- N — Existing amount at this time
- N_0 — Initial amount
- r — Percentage change per day/hour/year
- n — Number of days/hrs/yrs

Percentage Increase and Decrease

The (1+ r/100) bit might look a bit confusing in the formula but in practise it's really easy:

E.g 5% increase will be 1.05   5% decrease will be 0.95 (= 1 – 0.05)
26% increase will be 1.26   26% decrease will be 0.74 (= 1 – 0.26)

3 Examples to show you how EASY it is:

1) "A man invests £1000 in a savings account which pays 8% per annum. How much will there be after 6 years?"

ANSWER: Usual formula (as above): Amount = $1000(1.08)^6$ = £1586.87

(1000 = Initial amount, 1.08 = 8% increase, 6 = 6 years)

2) "The activity of a radio-isotope falls by 12% every hour. If the initial activity is 800 counts per minute, what will it be after 7 hours?"

ANSWER: Same old formula:

Activity = Initial value$(1 - 12/100)^n$

Activity = $800(1 - 0.12)^7 = 800 \times (0.88)^7$ = 327 cpm

3) "In a sample of bacteria, there are initially 500 cells and they increase in number by 15% each day. Find the formula relating the number of cells, n and the number of days, d."

ANSWER: Well stone me, it's the same old easy-peasy compound increase formula again: $n = n_0(1+0.15)^d$ or finished off: $n = 500 \times (1.15)^d$

The Acid Test:

LEARN THE FORMULA. Also learn the 3 Examples. Then turn over and write it all down.

1) A colony of stick insects increases by 4% per week. Initially there are 30. How many will there be after 12 weeks?
2) The speed of a tennis ball rolled along a smooth floor falls by 16% every second. If the initial speed was 5m/s find the speed after 20 seconds. How long will it take to stop?

Direct and Inverse Proportion

Direct Proportion: $y = kx$

BOTH INCREASE TOGETHER

1) The graph of y against x is a straight line through the origin: $y = kx$

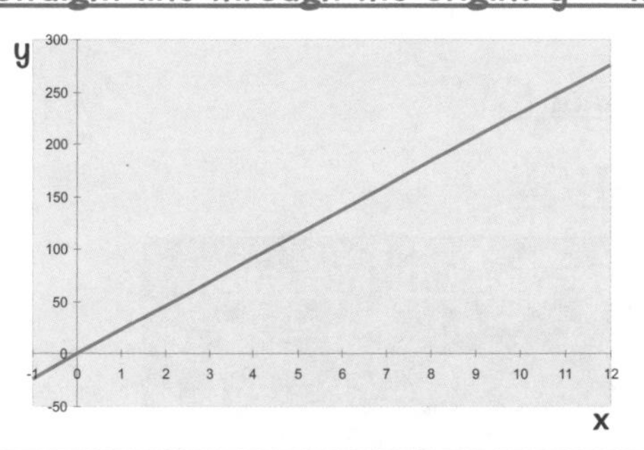

2) In a table of values the MULTIPLIER is the same for X and Y. i.e. if you double one of them, you double the other; if you times one of them by 3, you times the other by 3, etc.

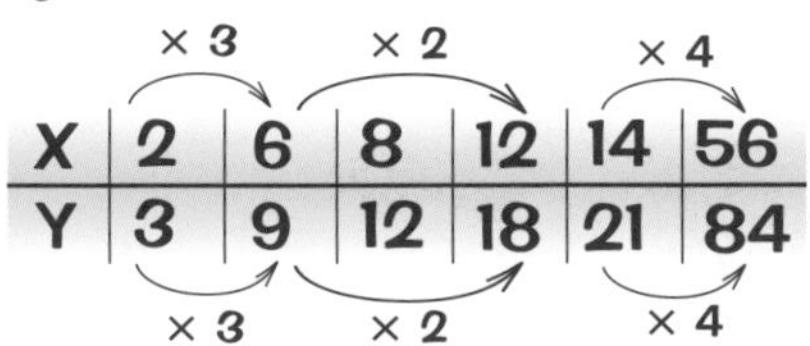

X	2	6	8	12	14	56
Y	3	9	12	18	21	84

(X: 2 → 6 × 3, 6 → 12 × 2, 14 → 56 × 4; Y: 3 → 9 × 3, 9 → 18 × 2, 21 → 84 × 4)

3) The RATIO x/y is the same for all pairs of values, i.e from the table above:

$$\frac{2}{3} = \frac{6}{9} = \frac{8}{12} = \frac{12}{18} = \frac{14}{21} = \frac{56}{84} = 0.6667$$

Inverse Proportion: $y = k/x$

One INCREASES, one DECREASES

The graph of y against x is the well known $y=k/x$ graph:

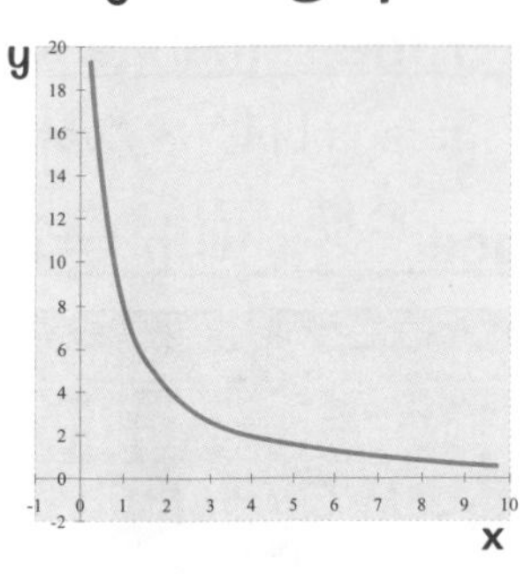

In a table of values the MULTIPLIER for one of them becomes a DIVIDER for the other, i.e. if you double one, you half the other, if you treble one, you divide the other by three, etc.

X	2	6	8	12	40	10
Y	30	10	7.5	5	1.5	6

(X: 2 → 6 × 3, 6 → 12 × 2, 40 → 10 ÷ 4; Y: 30 → 10 ÷ 3, 10 → 5 ÷ 2, 1.5 → 6 × 4)

The PRODUCT XY (X times Y) is the same for all pairs of values,

i.e. in the table above

$2 \times 30 = 6 \times 10 = 8 \times 7.5 = 12 \times 5$
$= 40 \times 1.5 = 10 \times 6 = 60$

Inverse Square Variation

You can have all sorts of relationships between x and y, like $y = kx^2$ or $y = k/x^3$ etc. as detailed on the next page. The most important type is $y = k/x^2$ and is called "INVERSE SQUARE" variation.

DON'T MIX UP THIS NAME with inverse proportion, which is just $y = k/x$.

The Acid Test:

LEARN the 3 KEY FEATURES for both Direct and Inverse proportion. Then turn over and write them all down.

1) Give examples of 2 real quantities that exhibit a) direct- and b) inverse proportion.
2) Make up your own tables of values which show
a) DIRECT PROPORTION b) INVERSE PROPORTION

Variation

This page shows you how to deal with questions which involve statements like these:

"y is proportional to the square of x" "t is proportional to the square root of h"

"D varies with the cube of t" "V is inversely proportional to r cubed"

To deal successfully with things like this you must remember this method:

Method:

1) Convert the sentence into a proportionality
 using the symbol " $\propto$ " which means "is proportional to"

" $\propto$ " with "=k" to make an **EQUATION**:

ove examples would become:	Proportionality	Equation
oportional to the square of x"	$y \propto x^2$	$y = kx^2$
oportional to the square root of h"	$t \propto \sqrt{h}$	$t = k\sqrt{h}$
es with the cube of t"	$D \propto t^3$	$D = kt^3$
versely proportional to r cubed"	$V \propto 1/r^3$	$V = k/r^3$

nce you've got it in the form of an equation with k, the rest is easy)

AIR OF VALUES of x and y somewhere in the question, and
TE them into the equation with the sole purpose of finding k.

value of k back into the equation
ow ready to use, e.g. $y = 3x^2$

BLY, they'll ask you to find y,
given you a value for x (or vice versa).

e:

n for a duck to fall down a chimney (it happens!) is inversely proportional
of the diameter of the flue. If she took 25 seconds to descend a chimney
0.3m, how long would it take her to get down one of 0.2m diameter?

e's no mention of "writing an equation" or "finding k" — it's up to YOU to remember
for yourself)

ANSWER:

1) Write it as a proportionality, then an equation: $t \propto 1/d^2$ i.e. $t = k/d^2$
2) Sub in the given values for the two variables: $25 = k/0.3^2$
3) Rearrange the equation to find k: $k = 25 \times 0.3^2 = 2.25$
4) Put k back in the formula: $t = 2.25/d^2$
5) Sub in new value for d: $t = 2.25/0.2^2 = 56.25$ secs

The Acid Test:

LEARN the FIVE STEPS of the METHOD plus the four examples. Then turn over and write them all down.

1) The frequency of a pendulum is inversely proportional to the square root of its length. If the pendulum swings with a frequency of 0.5 Hz when the length is 80cm, what frequency will it have with a length of 50cm, and what length will give a frequency of 0.7 Hz?

Unit Four Revision Summary

Here we are again — more great questions to test yourself with.
Keep practising these questions over and over again until you can answer them all. Seriously, it's the best revision you can do.

Keep learning these basic facts until you know them

1) Name with examples the different forms that rational and irrational numbers can take.
2) What is a reciprocal and why does zero have no reciprocal?
3) How do you simplify awkward ratios, e.g. 3.25 : 4.5?
4) How do you split a total amount in a certain ratio?
5) Demonstrate the two methods for turning recurring decimals into fractions.
6) Do your own examples to illustrate each of the three types of percentage question.
7) What is the formula for percentage change? Give two examples of its use.
8) Write down the three steps needed to find the area of a segment.
9) What are the six rules for circle geometry?
10) Write down the volume formula for a) a sphere b) a prism c) a cone d) a pyramid.
11) What is the formula for the surface area of a) a sphere b) a cone c) a cylinder?
12) How can you tell if a formula is a length, area or volume just by looking at it?
13) Describe the different features of a distance-time graph and a velocity-time graph.
14) What is a locus? Describe, with diagrams, the four you should know.
15) Demonstrate how to draw accurate 60° and 90° angles.
16) What do "congruent" and "similar" mean? How do you tell if 2 triangles are congruent?
17) What three types of scale factor are there and what is the result of each?
18) What does TERRY stand for? What details must be given for each transformation type?
19) What is the formula for Pythagoras' theorem? Where can you use Pythagoras?
20) What are the three key words for bearings? How must bearings be written?

Sorry, you're not finished yet... there's more over the page.

Unit Four Revision Summary

21) Write down the steps of a good, solid method for doing TRIG.

22) Write down the SINE and COSINE rules and draw a properly labelled diagram.

23) What's the main rule for adding vectors?

24) Do the "Queen Mary's tugboats" question using SINE and COSINE rules.

25) What three steps allow you to find the angle between a line and a plane?

26) What do "m" and "c" in $y = mx + c$ represent?

27) How are the gradients of perpendicular and parallel lines related?

28) List the five "harder" types of graph that you might have to find an equation for.

29) Draw the graphs of SIN, COS and TAN over 0° to 360° and then –1080° to 1080°.

30) Illustrate each of the four different types of graph transformation.

31) What is the rule for deciding what the area under a graph actually represents?

32) What is the rule for deciding what the gradient of a graph represents?

33) What are the 2 steps for identifying "a and b" in an equation from its graph?

34) What are the seven steps for solving equations or rearranging formulas?

35) What do you need to do if you divide an inequality by a negative number?

36) What is the four-stage method for graphical inequalities?

37) Write down the quadratic formula.

38) What are the four main steps for turning a quadratic into a completed square?

39) Write down the four steps of the trial and improvement method.

40) Write down the six steps for easy and seven steps for tricky simultaneous equations.

41) What is the formula for compound growth and decay? Give 3 examples of its use.

42) Give an example of direct proportion and an example of inverse proportion.

43) What are the five steps for dealing with questions on variation?

Answers

Unit Two — Acid Tests

P.2 Probability:

1) a) Probability = QQA+QAQ+AQQ = (4/52)(3/51)(1/50) + (4/52)(1/51)(3/50) + (1/52)(4/51)(3/50) = 3/11050

b) Probability = (4/52) × (3/51) + (4/52) × (3/51) + (4/52) × (3/51) = 3 × (4/52) × (3/51) = 3/221

P.3 Relative Frequency:

1) Probability of Bill drawing an ace = (13/100) = 0.13 Expected probability of drawing an ace = (4/52) = 0.077. 0.13 is much greater than 0.077 so YES, the pack is biased.

P.4 Probability — Tree Diagrams: **2)** 8/15

P.5 Probability — Tree Diagrams:

1) (2/7) × (1/6) × (5/6) × (4/5) = (40/1260) = (2/63)

P.6 Mean, Median, Mode and Range:

First: -14, -12, -5, -5, 0, 1, 3, 6, 7, 8, 10, 14, 18, 23, 25

Mean = 5.27, Median = 6, Mode = -5, Range = 39

P.8 Frequency Tables:

No. of Phones	0	1	2	3	4	5	6	TOTALS
Frequency	1	25	53	34	22	5	1	141
No. × Frequency	0	25	106	102	88	25	6	352

Mean = 2.5, Median = 2, Mode = 2, Range = 6

P.9 Grouped Frequency Tables:

Length L (cm)	15.5≤L<16.5	16.5≤L<17.5	17.5≤L<18.5	18.5≤L<19.5	TOTALS
Frequency	12	18	23	8	61
Mid-Interval Value	16	17	18	19	—
Freq × MI V	192	306	414	152	1064

Mean = 17.4, Modal Group = 17.5 ≤ L < 18.5, Median ≈ 17.5

P.10 Cumulative Frequency:

Median = 58 kg,
Lower Quartile = 53 kg
Upper Quartile = 62 kg,
Inter-quartile range = 9 kg

No. of fish	41 – 45	46 – 50	51 – 55	56 – 60	61 – 65	66 – 70	71 – 75
Frequency	2	7	17	25	19	8	2
Cum. Freq.	2	9	26	51	70	78	80

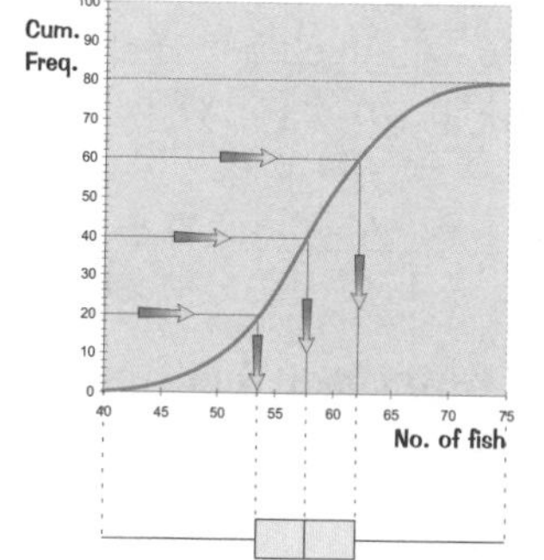

P.11 Histograms and Frequency Density:

1) 0–5: 9, 5–10: 27, 10–15: 36, 15–20: 45, 20–25: 27, 25–35: 18, 35–55: 18, 55–65: 36, 65–80: 162, 80–90: 126, 90–100: 18

2)
0 | 3 6 7
1 | 1 2 3 4 6 7 9
2 | 0 2 4 6 6

P.12 Spread of Data:

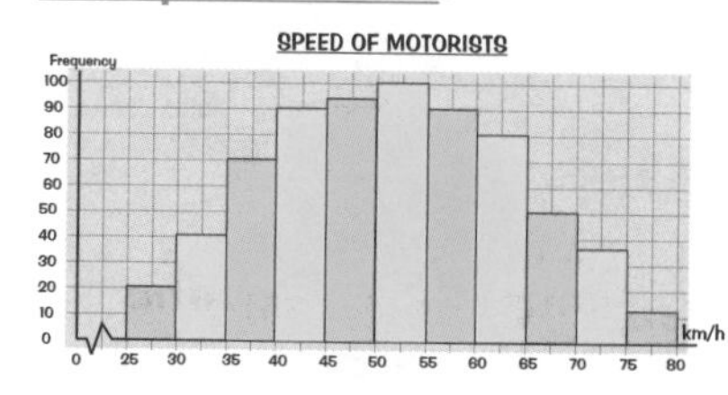

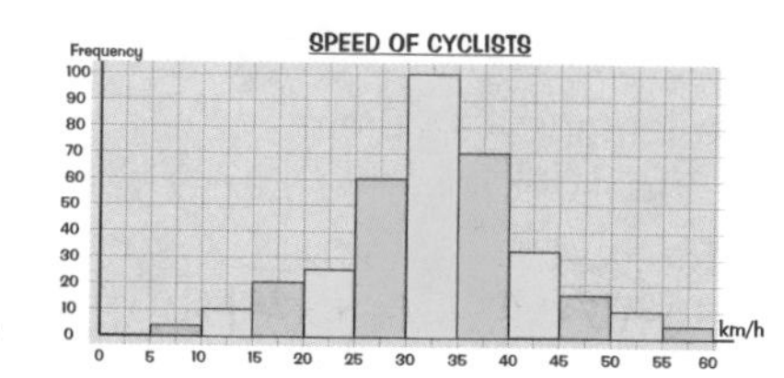

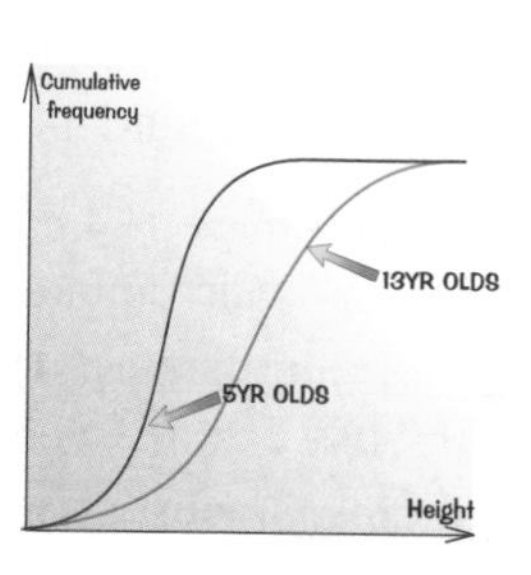

P.13 Other Graphs and Charts:

2) Guinea Pigs 68°, Rabbits 60°, Ducks 104°, Stick insects 48°

P.14 Sampling Methods:

1) Sample too small, motorways not representative of average motorist, only done at one time of day and in one place. Better approach: Take samples from a range of different locations across the country, take samples at different times of day, have a much larger sample size, e.g. 1000.

P.15 Time Series:

1) a) period = 4 months **b)** Find the average of the readings from months 1-4, then the average from months 2-5, then from months 3-6, etc. (& you could plot these on a graph to see the trend.)

Answers

Unit Three — Acid Tests

P.17 Types of Number: **1)** Squares: 256, 289, 324, 361, 400 Cubes: 1331, 1728, 2197, 2744, 3375 Triangles: 66, 78, 91, 105, 120 Powers of 2: 64, 128, 256, 512, 1024 Powers of 10: 1000 000, 10 000 000, 100 000 000, 1000 000 000, ... Primes: 47, 53, 59, 61, 67 **2) a)** n^2, n^3, $\frac{1}{2}n(n+1)$

P.18 Prime Numbers: **1)** 2, 3, 5, 7, 11, 13, 17, 19, 23, 29, 31, 37, 41, 43, 47 **2) a)** 101, 103, 107, 109 **b)** none **c)** 503, 509

P.19 Multiples, Factors and Prime Factors: **1)** 7, 14, 21, 28, 35, 42, 49, 56, 63, 70 ; 9, 18, 27, 36, 45, 54, 63, 72, 81, 90 ; LCM=63 **2)** 1, 2, 3, 4, 6, 9, 12, 18, 36 ; 1, 2, 3, 4, 6, 7, 12, 14, 21, 28, 42, 84 ; HCF=12 **3) a)** $990=2\times3\times3\times5\times11$ **b)** $160=2\times2\times2\times2\times2\times5$

P.20 Fractions, Decimals and Percentages: **1) a)** 6/10 = 3/5 **b)** 2/100 = 1/50 **c)** 77/100 **d)** 555/1000 = 111/200 **e)** 56/10 = 28/5

P.21 Finding the nth Term: **1) a)** $3n + 1$ **b)** $5n - 2$ **c)** $\frac{1}{2}n(n+1)$ **d)** $n^2 - 2n + 4$

P.23 Calculator Buttons: **1)** a) 11/4 **b)** 33/2 **c)** 33/4

3) (23.3 + 35.8) ÷ (36 × 26.5) =

4)a) 4hrs 34mins (12secs) **b)** 5.5397 hrs

P.24 Conversion Factors: **1)** 2,300m **2)** £34 **3)** 3.2cm

P.25 Metric and Imperial Units: **1)** 16 litres **2)** 200 or 220 yards **3)**115 cm **4)** £1.09 per litre **5)**104 km/h

P.26 Rounding Numbers: **1)** 3.57 **2)** 0.05 **3)** 12.910 **4)** 3546.1

P.27 Rounding Numbers: **1) a)** 3.41 **b)** 1.05 **c)** 0.07 **d)** 3.60 **2 a)** 568 (Rule3) **b)** 23400 (Rule 3) **c)** 0.0456 (Rules 1 and 3) **d)** 0.909 (Rules 1, 2 and 3) **3)** 16 feet 6 in. up to 17 feet 6 in

P.28 Basic Algebra: **1) a)** +12 **b)** -6 **c)** x **d)** -3
2)a) +18 **b)** -216 **c)** 2 **d)** -27 **e)** -336
3)a) $(x - 4y)(x + 4y)$ **b)** $(7 - 9pq)(7 + 9pq)$ **c)** $3(2yx^3 - 4k^2m^4)(2yx^3 + 4k^2m^4) = 12(yx^3 - 2k^2m^4)(yx^3 + 2k^2m^4)$

P.30 Basic Algebra: **1)** $4x + y - 4$ **2)** $6p^2q - 8pq^3$
3) $8g^2 + 16g - 10$ **4)** $7xy^2(2xy + 3 - 5x^2y^2)$ **5)** $c^4/6d^3$ **6)** $\dfrac{2(17g - 6)}{5(3g - 4)}$

P.31 Powers and Roots: **1)a)** 3^8 **b)** 4 **c)** 8^{12} **d)** 1 **e)** 7^6 **2)a)** 64 **b)** 1/625 **c)** 1/5 **d)** 2 **e)** 125 **f)** 1/25 **3)a)** 1.53×10^{17} **b)** 15.9 **c)** 2.89

P.33 Standard Index Form: **1)** 8.54×10^5 ; 1.8×10^{-4} **2)** 0.00456 ; 270,000 **3)a)** 2×10^{11} **b)**1×10^8 **4)** 6.5×10^{102}

P.34 Areas, Solids and Nets: **1)** 128.8 cm^2 **2)** 294 cm^2 **3)** 174 cm^2 **4)** 96 cm^2

P.35 Volume or Capacity:
1) a) Trapezoidal Prism, V = 148.5 cm^3 **b)** Cylinder, V = 0.70 m^3

P.37 Geometry: **1)** 68^0 and 44^0, or both 56^0 **2)** $x=66^0$

P.38 Circles: **1)** Area = 154 cm^2 Circumference = 44.0 cm
2) Area = 113 m^2 Circumference = 37.7 m

P.39 X, Y and Z Coordinates: **1)** A(4,5) B(6,0) C(5,-5) D(0,-3) E(-5,-2) F(-4,0) G(-3,3) H(0,5) **2) a)** (5, 2.5) **b)** (-2, 2.5) **c)** (0, -3.5) **3)** A(7,0,0), C(0, 4, 0), D(0, 0, 0), E(7, 0, 2), G(0, 4, 2), H(0, 0, 2)

P.41 Straight Line Graphs: **1)**

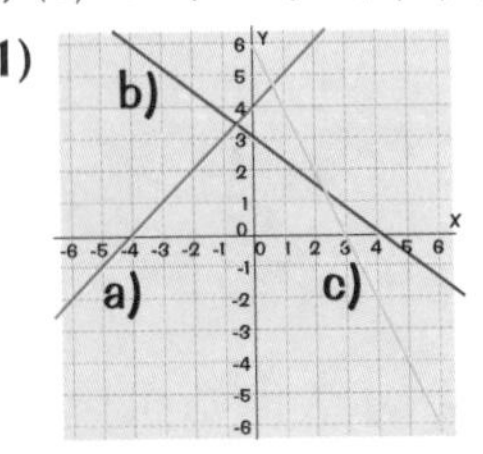

P.42 Formula Triangles: **1)** 16.5g/cm^3 **2)** 602.7g **3) a)** F / m×a **b)** F / P×A **c)** V / I×R

P.43 Speed, Distance and Time: **1)** 1 hr 37 mins 30 secs **2)** 1.89km = 1890m

P.44 Accuracy and Estimating: **1) a)** 35g **b)**134 mph **c)** 850g
2) a) Approx 600 miles × 150 miles = 90,000 square miles **b)** Approx 7cm × 7cm × 10cm = 490cm^3
3) a) 3.4 or 3.5 **b)** 10.1 or 10.2 **c)** 7.1, 7.2 or 7.3 **d)** 5.4 or 5.5

Answers

Unit Four — Acid Tests

P.46 Manipulating Surds and Use of π: **1)** $4\sqrt{2}$ **2)** $1 + 2\sqrt{2}$

P.47 Upper bounds and Reciprocals: **1 a)** x — lower bound 2.315 m, upper bound 2.325 m y — lower bound 0.445 m, upper bound 0.455 m **b)** max value of z = 4.57, min value of z = 4.51

P.49 Ratios: **1a)** 5:7 **b)** 2:3 **c)** 3:5 **2)** 17½ bowls of porridge **3)** £3500 : £2100 : £2800

P.50 Fractions: **1) a)** 5/32 **b)** 32/35 **c)** 23/20 **d)** 1/40 **e)** 167/27 **2) a)** 220 **b)** £1.75

P.51 Fractions: **1)** 1/7 **2)** 7/200

P.52 Percentages: **1)** 40% **2)** £20,500 **3)** 1.39%

P.53 Regular Polygons: **1)** 72^0, 108^0 **2)** 30^0, 150^0 **3)** 15 sides
4) Pentagon, all the angles are either 36^0, 72^0 or 108^0 Octagon as shown →

P.55 Circles: **2)** BCD = 90°, CBO = 42°, OBE = 48°, BOE = 84°, OEF = 90°, AEB = 42°, OEB = 48°

P.56 Volumes: **1)** Cone, 20.3 m^3 **2)** 33.5 cm^3, 179.6 cm^3

P.57 Surface Area and Projections: **1)** 50.3 cm^2 and 153.9 cm^2 **2)** 364 cm^2

P.58 Length, Area & Volume: 1) πr^2 = Area, Lwh = Volume, πd = Perimeter, ½bh = Area, $2bh + 4l\pi$ = Area, $4r^2h + 3\pi d^3$ = Volume, $2\pi r(3L + 5T)$ = Area **2) a)** 230,000cm^2 **b)** 3.45m^2
3) a) 5,200,000cm^3 **b)** 0.1m^3

P.59 D/T Graphs and V/T Graphs: **1)** 0.5km/h
2) Accels. 6m/s^2 2m/s^2 -8m/s^2(deceleration), Speeds: 30m/s 50m/s

P.63 Similarity and Enlargements: **1)** A'(-3,-1.5), B'(-7.5,-3), C'(-6,-6) **2)** 64m^2

P.64 The Four Transformations: **A→B** Rotation of 90° clockwise about the origin,
B→C Reflection in the line y = x, **C→A** Reflection in the y-axis, **A→D** Translation of $\binom{-9}{-7}$

P.65 Combinations of Transformations:
1) **C→D**, Reflection in the Y-axis, and an enlargement SF 2, centre the origin
   D→C, Reflection in the Y-axis, and an enlargement SF ½, centre the origin.
2) **A'→B**, Rotation of 180° clockwise or anticlockwise about the point (0,3).

P.66 Symmetry:
1) H : 2 lines of symmetry, Rot^{nl}. symmetry Order 2,
   E : 1 line of symmetry, no Rotational symmetry,
   M : 1 line of symmetry, no Rotational symmetry,
   S : 0 lines of symmetry, Rot^{nl}. symmetry Order 2,
   N: 0 lines of symmetry, Rot^{nl}. symmetry Order 2
   Y: 1 line of symmetry, no Rot^{nl}. symmetry
   B: 1 line of symmetry, no Rot^{nl}. symmetry
   T: 1 line of symmetry, no Rot^{nl}. symmetry
2) Left to right, no of planes of symmetry are: 9 (this one is the hardest, by the way), 3, 4, ∞, 4

P.67 Pythagoras' Theorem and Bearings: **1)** BC=8m **2)** 298^0 **3)** 118^0

P.69 Trigonometry — Sin, Cos, Tan: **1)** x=26.5 m **2)** 23.6^0 **3)** 32.6^0 (both)

P.71 The Sine and Cosine Rules: **2)** 17.13m, 68.8^0, 41.2^0

P.72 Vectors:

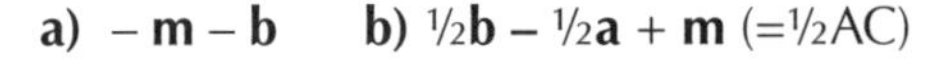

a) − **m** − **b** **b)** ½**b** − ½**a** + **m** (=½AC)

c) ½(**a** − **b**) + **m** **d)** ½(**b** − **a**)

P.73 "Real Life" Vector Questions: Resultant Force: R = 21800 N, θ = 3.4°

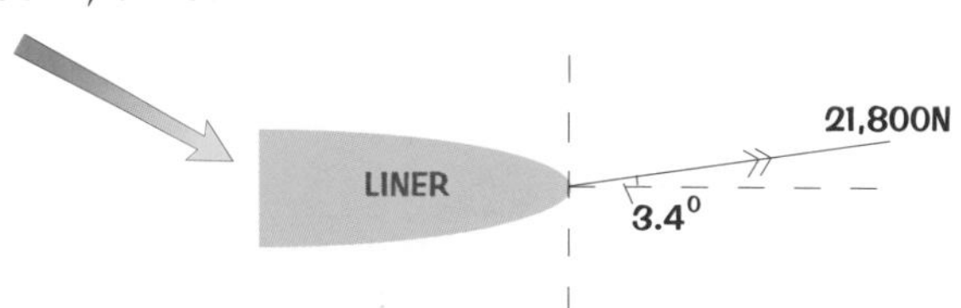

P.74 3-D Pythagoras and Trigonometry:

1) 25.1° **2)** 7.07 cm

P.75 Straight Line Graphs: Gradient: **1)** -1.5

P.76 Straight Line Graphs: y = mx+c:

1)

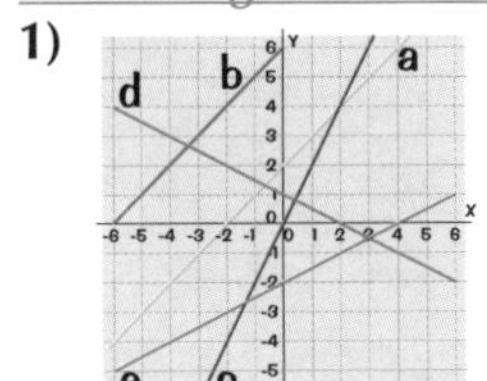

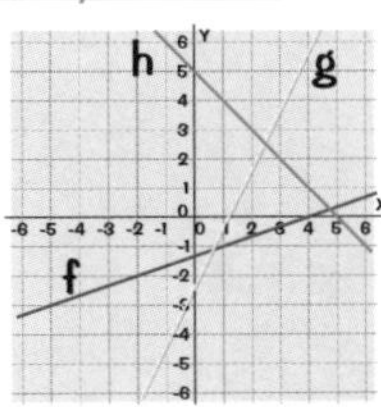

Answers

P.77 Pythagoras, Lines and Line Segments:
1) 5 units **2)** A line continues to infinity, whereas a segment is part of a line.

P.78 Some Harder Graphs to Learn: **1)a)** x^2 bucket shape **b)** $-x^3$ wiggle (top left to bottom right)
c) +ve inverse proportion graph **d)** circle about origin, radius 6 **e)** –ve inverse proportion graph
f) $+x^3$ wiggle (bottom left to top right) **g)** $-x^2$ upside down bucket shape **h)** K^X curve upwards through (0,1)

P.79 Quadratic graphs:
See graph to the right. Using graph, solutions are x = -2 and x = 3.

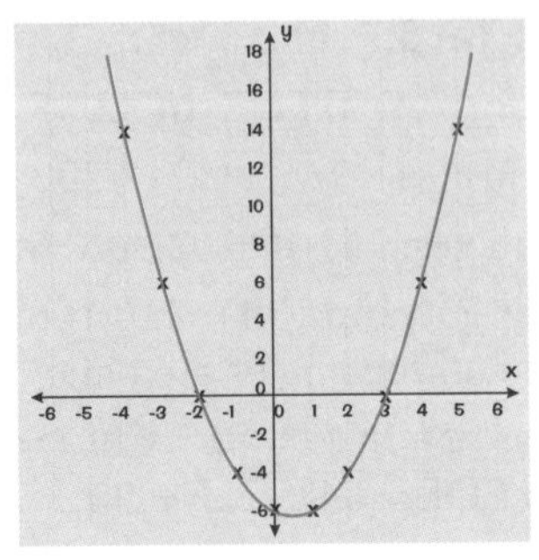

P.81 Angles of Any Size:

1)a) $x = -330^0, -210^0, 30^0, 150^0, 390^0, 510^0$
b) $x = -228^0, -132^0, 132^0, 228^0, 492^0, 588^0$
c) $x = -315^0, -135^0, 45^0, 225^0, 405^0, 585^0$

P.83 Graphs: Shifts and Stretches:

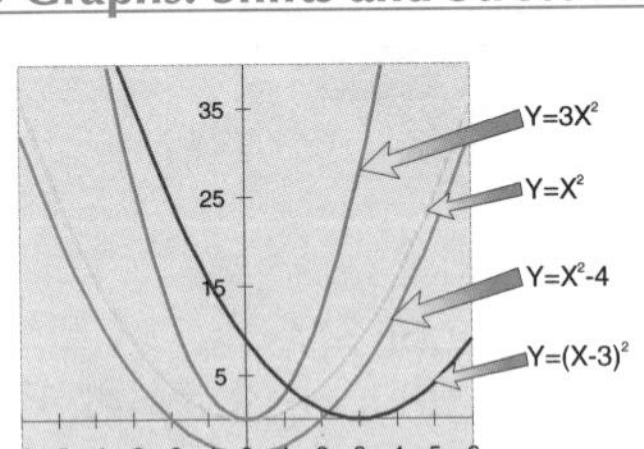

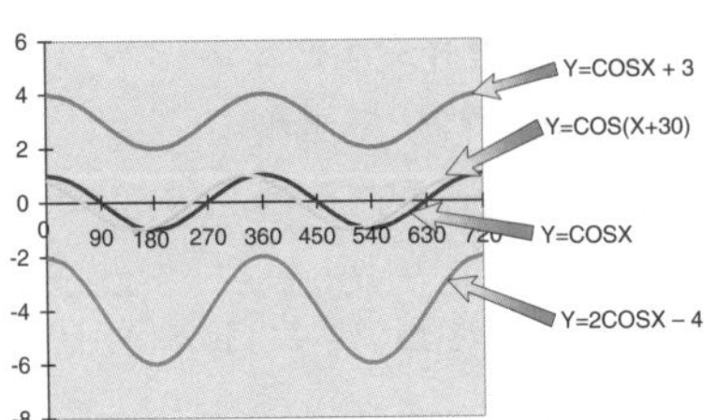

P.84 The Meaning of Area and Gradient: **1)** Total babies born **2)** Fuel consumption in "miles per gallon"

P.85 Finding Equations from Graphs: **1)** P = 50, Q = 1.4

P.86 Solving Equations: **1)a)** x=2 **b)** x = - 0.2 or -1/5 **c)** x = ±3

P.87 Rearranging Formulas: **1)** C=5(F–32)/9 **2)a)** p=-4y/3 **b)** p=rq/(r+q) **c)** p=±√{rq/(r+q)}

P.88 Inequalities: **1)** $-2 \leqslant x$ **2)a)** -6, -5, -4, -3, -2, -1, 0, 1, 2, 3, 4, 5, 6 **b)** -4, -3, -2, -1, 0, 1, 2, 3, 4

P.89 Graphical Inequalities:

1)

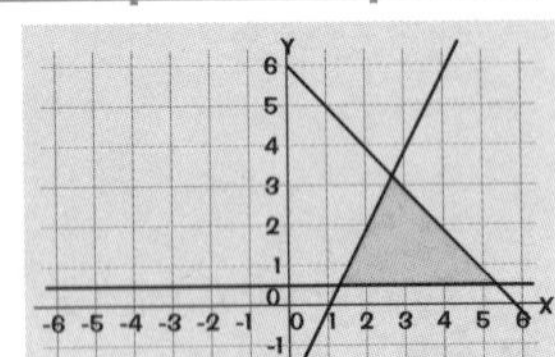

P.91 Factorising Quadratics: **1a)** x = 3 or -8 **b)** x=7 or -1 **c)** x = 1 or -7 **d)** x = 4 or -3/5

P.92 The Quadratic Formula: **1)a)** x = 0.39 or -10.39 **b)** x = 1.46 or -0.46 **c)** x = 0.44 or -3.44

P.93 Completing the Square: **2)a)** x = 0.39 or -10.39 **b)** x = 1.46 or -0.46 **c)** x = 0.44 or -3.44

P.94 Trial and Improvement: **1)** x = 2.4

P.95 Simultaneous Equations and Graphs:
1) a) x = 2, y = 4 **b)** x = 1½, y = 3
2) Solutions are: x = -2 or x = 1.5

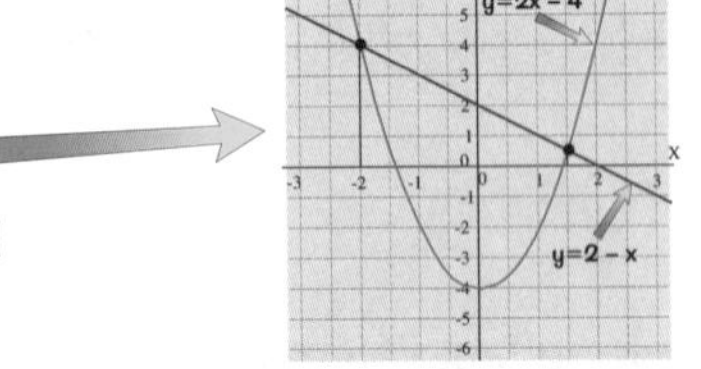

P.96 Simultaneous Equations: **2)** F=3, G=-1

P.97 Simultaneous Equations:
a) f = 4 & g = 0 OR f = 40 & g = 6
b) f = -5/3 & g = -2/3 OR f = 72 & g = 5 **c)** f = 9 & g = -3/2 OR f = 1 & g = 1/2
d) f = -3 & g = 33 OR f = 1/4 & g = -11/4

P.98 Compound Growth and Decay: **1)** 48 stick insects **2)** 0.15m/s. Forever.

P.99 Direct and Inverse Proportion: **1)a)** Total cost vs No.of tins of Bone-tingling Fireball Soup
b) No. of people working on a job vs time taken to complete it.

P.100 Variation: **1)a)** 0.632 Hz **b)** 40.8 cm

Index

Index

MEHR42